Diarmid N. Patron

Report of Investigations on the Life-History of the Salmon in Fresh Water

from the Research laboratory of the Royal College of Physicians of Edinburgh

Diarmid N. Patron

Report of Investigations on the Life-History of the Salmon in Fresh Water
from the Research laboratory of the Royal College of Physicians of Edinburgh

ISBN/EAN: 9783337094133

Printed in Europe, USA, Canada, Australia, Japan

Cover: Foto ©berggeist007 / pixelio.de

More available books at **www.hansebooks.com**

FISHERY BOARD FOR SCOTLAND.

SALMON FISHERIES.

REPORT

OF

INVESTIGATIONS

ON THE

LIFE HISTORY OF SALMON.

Presented to Parliament by Command of Her Majesty.

GLASGOW :
PRINTED FOR HER MAJESTY'S STATIONERY OFFICE
By JAMES HEDDERWICK & SONS,
At "The Citizen" Press, St. Vincent Place.

And to be purchased either directly or through any Bookseller, from
JOHN MENZIES & CO., 12 Hanover Street, Edinburgh, and
90 West Nile Street, Glasgow ; or
EYRE & SPOTTISWOODE, East Harding Street, Fleet Street, E.C.; or
HODGES, FIGGIS & CO., Limited, 104 Grafton Street, Dublin.

1898.

[C.—8787.] Price 1s. 11d.

FISHERY BOARD FOR SCOTLAND,
EDINBURGH, *23rd February*, 1898.

SIR,

I am directed by the Fishery Board for Scotland to transmit herewith a Report of Investigations into the Life-History of Salmon in Fresh Water, which was submitted to the Board at a meeting held on 18th instant, and ordered by them to be forwarded to you, for the information of the Secretary for Scotland.

I have the honour to be,

SIR,

Your most obedient Servant,

WM. C. ROBERTSON,
Secretary.

The Under-Secretary for Scotland.

REPORT OF INVESTIGATIONS

ON THE

LIFE-HISTORY

OF THE

SALMON IN FRESH WATER:

FROM

THE RESEARCH LABORATORY OF THE ROYAL COLLEGE

OF PHYSICIANS OF EDINBURGH.

EDITED BY

D. NOËL PATON, M.D.,

SUPERINTENDENT OF THE LABORATORY.

GLASGOW:
PRINTED BY JAMES HEDDERWICK & SONS,
AT "THE CITIZEN" PRESS, ST. VINCENT PLACE.

1898.

RESEARCH LABORATORY OF THE ROYAL COLLEGE OF
PHYSICIANS OF EDINBURGH,
January, 1898.

To THE SCOTTISH FISHERY BOARD,

GENTLEMEN,—

I have the honour to present the following Report
of the Investigations on the Life-History of the Salmon in Fresh
Water, which have been carried on in the Research Laboratory of
the Royal College of Physicians of Edinburgh.

In doing so, I desire to state that the valuable plan of procuring
salmon simultaneously from the mouths and from the upper reaches
of the rivers for the investigation of the migrations of the fish and
the elucidation of the question of whether fish leaving the sea early
in the year remain in the upper waters until the breeding season,
is due to Mr. Walter E. Archer, Inspector of Salmon Fisheries for
Scotland.

On behalf of my fellow-workers and myself, I take this opportunity of thanking him for much valuable advice and assistance
during the prosecution of the work.

I have the honour to be, Gentlemen,

Your obedient Servant,

D. NOËL PATON,

Superintendent of the Laboratory of the Royal
College of Physicians of Edinburgh.

CONTENTS.

INVESTIGATIONS ON THE LIFE-HISTORY OF THE SALMON IN FRESH-WATER.

I.—INTRODUCTORY.

1.—GENERAL INTRODUCTION.

By D. NOËL PATON, M.D., F.R.C.P.Ed., B.Sc.

SUPERINTENDENT OF THE LABORATORY.

The curious life history of the salmon has always been a subject of the deepest interest not only to the zoologist and physiologist but also to the sportsman and the fisherman. In spite of the most careful study by scientific investigators, the migrations of the salmon and the various changes in condition which it undergoes are even now far from being fully understood, and the careless observations and foolish traditions of keepers, fishermen, and ghillies have only served to involve the matter in a deeper cloud of mystery.

Only a few years ago the processes of reproduction and development were matters of speculation, and many of the older writers indulged in the most fanciful ideas upon these points. The investigations and experiments of Sir James Maitland and others have supplied the required information, and the reproduction and development of the salmon are no longer mysteries.

But many other questions in its life history remain unsolved. Yearly, or at longer intervals, the fish appear on our coast, apparently from the deeper waters, and ascend the rivers, there, somewhere between October and January. to deposit their spawn and milt. Having done so they descend the river as "kelts," and again disappear in the sea, to return either in the same or in the following year to the fresh water.

What force urges the fish to leave its rich feeding ground in the sea? Is it necessary that it should enter fresh water in order to perform the act of reproduction? Does it require or procure any food during its sojourn in the river, and, if not, how is it able to maintain life and to construct its rapidly growing genital organs?

In the female the growth of these is enormous. In April or May the ovaries constitute only about 1·2 per cent. of the weight of the fish—in November they are no less than 23·3 per cent. In a fish of 30lbs. in the spring they weigh about 120 grms.—in November they weigh over 2000 grms. The increase in the testes in the male is not so marked, but is sufficiently striking. In April or May these organs are about 0·15 per cent. of the weight of the fish, while in November they are 3·3 per cent.

From what are these structures formed? As they grow, the muscle, as is well known, undergoes marked and characteristic changes. Not only does it diminish in amount as the season advances, so that the fish which have been some time in the river become smaller in the shoulder and back, but it loses its rich, fatty character, while it becomes paler in colour.

Are these changes in the muscle connected with the growth of the ovaries and testes? And if so, in what manner and to what extent?

On the other hand, in fighting its way up rapids and over falls an enormous amount of muscular work is accomplished by the salmon. Whence is the energy for this work obtained? Are the changes in the muscle connected with the performance of this work, and if so to what extent are these changes connected with the muscular work, and to what extent with the growth of the genitalia. Lastly, the question arises to what extent do these changes in the muscle modify the value of the flesh as a food stuff.

In the investigation of some of these questions most excellent work has already been done, not only in Holland and Germany upon the salmon in the Rhine by Dr. Hoek[*] and Professor Miescher Ruesch,[†] but also by Mr. Archer, the Inspector of Salmon Fisheries for Scotland, in conjunction with Mr. Grey and Mr. Tosh. The careful series of observations embodied in the Annual Reports[‡] are well worth careful study by the zoologist and the salmon fisher. They should help to dispel the absurd traditions which cling around the history of the salmon, and to pave the way for the complete solution of many of the problems we have enumerated.

The present investigation is a continuation and amplification of these researches, and would have been impossible without these previous laborious studies.

Briefly stated, these investigations of the Fishery Board have established the following facts:—

1st. That some salmon spawn every year, though there is strong evidence that all do not do so. (*Eleventh Annual Report, Part II., p. 68.*)

2nd. That the genitalia of fish coming from the sea develop steadily from April on to the spawning time, and that the genitalia of salmon in the earlier summer months develop more rapidly than those of grilse. (*Fourteenth Annual Report, Part II., pp. 15 and 21.*)

3rd. That the proportion of the weight of genitalia to the weight of the fish is constant for all sizes of salmon. (*Fourteenth Annual Report, Part II., p. 11.*)

4th. That salmon continue to feed while in the sea until September. This is shown, firstly, by the presence of food in the stomach of a certain proportion of the fish captured (*Fourteenth Annual Report, Part II., pp. 77 to 80.*); and secondly, by the fact that the fish leaving the sea are somewhat heavier—from 2 to 3 per cent.—in August and September than they are in the earlier months, whereas if they had entirely stopped

* Rapport over Statistische en biologische onderyoekingen ingesteld mett behulp van in Nederland gevangen Zalmen. Dr. F. P. C. Hoek, Wetenschappelijk Adviseur in Visscherijzaken.
† Statistische und biologische Beiträge zur Kenntniss von Leben des Rheinlachses im Süsswasser Dr. F. Miescher Ruesch, Prof. d. Physiol. in Basel. A contribution to the literature of the Berlin Fisheries Exhibition of 1880. Publishers, von Metzger & Wittig, Leipsic. See also Histochemische und Physiol gischen Arbeiten von Friedrich Miescher. Bd II. s. 116. Leipzig, 1897.
‡ Appendices to Thirteenth and Fourteenth Annual Reports of the Fishery Board for Scotland, Part II.

feeding they should have been lighter.   (*Fourteenth Annual Report, Part II., p. 12.*)

If salmon do feed in the sea it is perhaps curious that food should be found in so small a percentage of those captured at the mouths of rivers.   But it must be remembered that the estuary of the river is not the natural feeding ground of the salmon, and it is probably only by chance that food is still in the stomach of fish captured there.

Scope of Present Investigation.

The present investigation was undertaken with the following objects : —

1. To throw light upon the influences determining the migration of salmon from sea to river and from river to sea, and to study more fully than has hitherto been done the course of the migrations.

2. To investigate whether salmon in fresh water require or use food; and whether fish which do not leave the sea till late in the autumn continue to feed while their genitalia are developing ?

3. To study the relative rate of growth of genitalia in salmon in fresh water and in the sea.

4. To investigate more fully the nature of the changes in the flesh (muscle) and genital organs throughout the year.

5. To determine the source of the material used in the construction of the genital organs, and to study the chemical changes which the various substances undergo.

6. To elucidate the source of the energy required by the fish in ascending the river.

Method of Investigation.

To carry out this enquiry, it was necessary to have a supply of fish at all possible seasons of the year.   Further, it was desirable to have fish from different sources at these various times—some from the sea at the mouth of the river, some from the upper reaches—so that the changes of the fish in their passage up the river, and in their sojourn in the upper waters, might be thoroughly studied.

It is known that through the spring, summer, and autumn there is a more or less constant procession or stream of salmon passing from the sea into the rivers.   Our plan, then, was to take samples of the fish just leaving the sea, and similar samples of those which had reached the upper parts of the rivers, and, by comparing the latter with the former, to investigate the changes which had taken place during the sojourn of the fish in fresh water.   The method may be compared to that of taking samples of water from two parts of a river in order, by examination of them, to ascertain what changes have taken place in the water between these points.

Material for Investigation.

(*a*) *Supply of Fish.*

The observations were commenced in June 1895 at the request of Mr. Archer, Inspector of Salmon Fisheries for Scotland.

At this time Mr. Tosh was stationed at Berwick-on-Tweed investigating the growth of the genitalia in Tweed salmon, and it was arranged that he should send the viscera, with portions of the flesh, to the Laboratory.

It was soon found that to make the investigation satisfactory it would

be necessary to determine the precise amount of muscle on the fish, and
since this was impossible in the fish required for market, Mr.
Johnston, of Montrose, was approached by Professor MacIntosh, of
St. Andrews, and, at his own expense, supplied a series of salmon caught
in his fishings on the North Esk throughout the season.

The material so accumulated and examined yielded results of considerable interest, but clearly showed that, to make the investigation
complete, a more extensive supply was necessary. Further, Mr.
Archer's Report to the Fishery Board for 1895 showed that in any such
investigation results obtained from grilse could not with safety be applied
to salmon. For this reason it was resolved to exclude grilse from
the investigation.

Mr. Archer threw himself into the progress of the enquiry with
enthusiasm, and, a grant having been obtained from the Fishery Board
to defray the expenses of collecting material, he made arrangements for a
large supply of fish throughout the season of 1896.

To thoroughly check and control observations, it was arranged to
procure salmon from different rivers, and the Helmsdale, Spey, Dee,
and Annan were selected.

It was further arranged that fish should be supplied first from the
mouth of each river, and second from the upper reaches.

Through the liberality of the various proprietors—the Duke of
Richmond, the Duke of Sutherland, the Duke of Fife, the Dee
(Aberdeenshire) District Board, Mr. Mackenzie of Newbie, and the
late Mr. Heywood-Lonsdale, the following supply of material was
obtained:—

1895.

From Mr. Johnston, of Montrose, 10 male and 9 female salmon and
grilse were received.

From Mr. Tosh, at Berwick-on-Tweed, and later at Melrose, the
viscera and a portion of the muscle of 18 fish were obtained.

Four whole salmon were late in the year received from the upper
waters of the Tweed through the courtesy of the Police Committee of
the Commissioners of the River.

1896.

On March 6th the viscera of seven salmon captured at the mouth of
the Tweed were procured at Berwick.

From May to November 69 salmon were received from the various
stations as follows:—

		May-June.	July-Aug.	Oct.-Nov.
Helmsdale,	Mouth,	3	2	1
	Upper,	3	4	5
Spey,	Mouth,	4	6	3
	Upper,	3	–	1
Dee,	Mouth,	5	5	5
	Upper,	3	5	3
Annan,	Mouth,	4	4	–
	Upper,	–	–	–
Totals,		25	26	18

In the spring of 1897, 22 kelts were received from the Spey.

In drawing conclusions from the study of this material one question must be considered. How far do fish captured in the upper waters actually represent the condition of all the fish there? These fish were for the most part taken with the fly, and it is the common opinion of anglers that fresh-run fish take the lure more readily than those which have been in the water for some time. It is thus possible that the fish procured from the upper waters do not fairly represent the general condition. Fortunately at Kincraig, on the Spey, the fish were taken with the net, and an examination of the various tables shows that they do not differ from the fish taken with the rod in the upper reaches of the other rivers.

(b) *Preparation and Preservation of Material.*

The duty of examining and preparing the fish as they were received in the Laboratory was discharged by Mr. Alfred Patterson, one of the assistants, who had had a special training in analytic chemistry, and his work was supervised by me.

The external appearance of the fish was observed, and the presence of sea lice, ulcers, wounds, &c., noted.

The fish was then measured as follows :— Length from mouth to fork of tail, depth at anterior border of dorsal fin, and girth at the same place. It was next weighed. The skin covering the trunk muscles on one side was next carefully removed, along with the anal fin of that side, any adherent flesh being afterwards carefully picked off and placed along with the rest. The skin was weighed.

The great trunk muscles on the same side were then carefully separated from the bones, and weighed. In some cases the " thin " belly muscle was separated from the " thick," and weighed separately. It was soon found that the " thin " was about a quarter of the whole muscle, and separate weighings were therefore discontinued.

Pieces of the " thick " and of the " thin," of about 30 grms. each, taken at the level of the anterior edge of the dorsal fin, were preserved in spirit. Another portion of the " thick " was weighed, and pounded up with an equal quantity of common salt.

The abdominal cavity was now opened, and the viscera, with the exception of the heart, the ovaries or testes, and the kidneys were removed and weighed. The liver was separated from the other viscera, and the condition of the gall bladder was noted. The gall bladder was removed and the liver weighed. A weighed portion of the organ— about 30 grms.— was put in methylated spirit.

In many cases inoculations upon gelatin were made from the stomach, pylorus, and intestine for the investigation of the bacteriology of the alimentary tract.

The stomach and intestine were opened, and the characters of the contents noted. Small pieces of stomach, intestine, pyloric appendages, and liver were in many cases placed in perchloride of mercury, and pieces of the liver were sometimes fixed in osmic acid for microscopic examination.

The stomach and œsophagus were cut off from the lower part of the alimentary canal, and were placed in alcohol. The pyloric appendages, with the intestine, were placed in alcohol in a separate bottle.

The ovaries or testes were now removed and weighed, the latter being weighed without the ducts, which were cut off close to the organs. Pieces of the organs, of about 30 grms., were preserved in methylated spirits.

The rest of the fish, consisting of the head, heart, vertebral column, kidneys, tail, fins, and the muscles of the other side, were weighed.

From these weighings **the total weight of** muscle and **the weight of**
"thick" and "thin" **could be calculated, as shown** by the **following**
example from the **Receiving-Book** :—

No. 23. Aberdeen, 27th May 1896.

From Mouth of the Dee.

Clean fish. **Sea** lice. Stomach empty. Intestine contained yellow
mucus and tape-worms. Pyloric appendage **coated** with **fat.** Gall
bladder distended.

Length, 67 cm.   Girth, 36·5 cm.   Depth, 15 cm.

Weight,	. . . . . .	.	3,425 grms.
„	of Skin on one side,	. .	110 grms.
„	of Viscera, ...	.	103 grms.
„	of Ovaries, ...	.	16 grms.
„	of Trunk Muscle of one side,	.	1,087 grms.
„	of Rest of Fish, . . .	.	2,005 grms.

The **weight of the trunk muscles was thus 2174 grms., and of this**
a **quarter, or 543, was "thin," and the remainder, or 1631, was**
"**thick.**"

(c) *Method of comparing Fish of different sizes.*

In **dealing with this mass of material the first** question which had to
be **considered was how to make the results** obtained from one fish
comparable with those obtained from another. The **fish** received varied
in length from 66 to 108 cm., and in weight from 2675 to 12,670 grms.
To compare the various parts of a fish of 3000 grms. with one of 10,000
grms. it is obviously necessary to reduce the results to some common
measure.

In previous observations the weight of the various organs has usually
been expressed simply as a percentage of the total weight of the fish ;
but, if the changes in substance of two structures such as the muscle and
ovaries—one of which steadily loses weight while the other steadily gains
it—are to be compared, obviously to calculate the changes in muscle, in
terms of the weight of the fish, will give a fallacious idea of the extent
of these changes.

A **more constant standard is the length of the** fish, and in his
Annual Report, No. 14, 2-12, Mr. Archer has used this standard.

But the weight of **the fish varies not as its length but** as some factor
of its length approaching **the cube, on the rule that two** bodies of similar
shape and of the same **material vary as the cube of their** length or some
like dimension. Thus a **fish of 100 cm.** in length is not twice as heavy
as a fish of 50 cm., but is eight times **as heavy.** On trial, in a large
number of cases of fish of different **sizes, it was found** that the cube
gave satisfactory results, and it **was therefore adopted as** the factor.

The standard **length, or unit of** length, **selected was** 100 cm.—the
ordinary length of a salmon of about 30lbs. This was selected simply
because it yielded **convenient** figures.

To obtain the weight per unit of length the proportion thus is :—

Actual length, cubed : standard length, cubed :: actual weight : *x*.

Table **I.** shows the correspondence of the **results** obtained in this
way with **the actual** weights observed in the very large number of fish
included in **Table** XIV., p. 27 of Vol. XIV. (1896) **of the** Fishery
Board Reports, **Part II.**

TABLE I.

Length.	Weight.					Cube Calculation.	Average of Actual Weights.
	May.	June.	July.	Aug.	Sept.		
75	4289	4133	3707	4533	4333	4219	4199
76 +	4324	4409	4503	4438	4653	4390 +	4469
77 +	4666	4577	4644	4817	4462	4565 +	4633
79 -	4706½	4982	4804	5018	5355	4930 -	4973
80	5160	5013	5249	5333	5129	5120	5184

I am indebted to Mr. Archer for the following observations on this point, made upon a very large series of fish of different lengths:—

TABLE II.

Length in Cm.	No. of Fish Examined.	Average Weight in Grms.	Weight for Fish of Standard Length.
70·0	102	3311·2	9716
79·0	265	4780·7	9777
88·0	106	6804·4	9684

The difference is thus less than 1 per cent.

METHOD OF INVESTIGATION.

It was impossible for a single individual to conduct so extensive an investigation as that contemplated, and I was fortunate in securing the co-operation of various workers in the Research Laboratory of the Royal College of Physicians.

The following scheme shows the general plan of the investigation and the manner in which it was apportioned among the various workers:—

How far may the Salmon examined be considered typical of their respective classes?—By Walter E. Archer, Inspector of Salmon Fisheries.

The Source from which Salmon obtain nourishment and the Exchange of Material in their Body during their sojourn in Fresh Water?—

A. *The Power of the Alimentary Canal to Digest and Absorb Food*

   1. Changes in the structure of the lining membrane of the alimentary canal and in the various glands. By G. Lovell Gulland.

Changes in the digestive activity of the various secretions of
the alimentary canal. By A. Lockhart Gillespie.
Bacteriology of the alimentary canal in different conditions.
By A. Lockhart Gillespie.

B. *Changes in the Weight and in the Condition of the Muscles, Genitalia,
and other Organs during the sojourn of the Salmon in Fresh
Water—*

Changes in the weight and condition of the fish at different
periods. By D. Noël Paton and J. C. Dunlop.
Changes in the amount of solids. By D. Noël Paton.
Changes in the amount of Fats—
    (a) Chemical observations. By D. Noël Paton.
    (b) Microscopic observations. By S. C. Mahalanobis.
The nature of the Proteids of the Muscle. By F. D. Boyd.
Changes in the amount of Proteids. By J. C. Dunlop.
The Fats and Proteids stored in the Muscle as the Source of
Muscle Energy. By D. Noël Paton.
The Phosphorus Compounds and the Changes in the distribu-
tion of Phosphorus. By D. Noël Paton.
Changes in the distribution of Iron. By E. D. W. Greig.
The Pigments of the Salmon and their Changes. By M. I.
Newbigin.
The Changes in the Value of the Salmon as a Food-Stuff.
By J. C. Dunlop.

2.—HOW FAR MAY THE SALMON EXAMINED BE CONSIDERED TYPICAL OF THEIR RESPECTIVE CLASSES?

By WALTER E. ARCHER,

INSPECTOR OF SALMON FISHERIES FOR SCOTLAND.

The object of these observations is to determine how far the fish examined in the course of this investigation may be taken to fairly represent fish frequenting each locality in the periods named.

The lower-water fish are represented by 34 females and 8 males, and upper-water fish by 21 females and 6 males. The lower-water fish were in each case taken at the mouths of the rivers. The fish from the upper waters of the Spey were taken at a distance of about 60 miles from the sea, those from the upper waters of the Dee at about 65 miles, and those from the upper waters of the Helmsdale at about 15 miles. Fish from 8 to 10 lbs. in weight were asked for, but those sent from the mouths in October and November were considerably larger. It has been shown that the *average* weight, per fish of standard length, of salmon measuring from 69 to 89 centimetres does not vary one per cent.—provided the averages are calculated over large numbers, and the fish are taken in the same locality and during the same period of the year (Table II., p. 7). It is well known, however, that there is a considerable variation in the weight of single fish of the same length, taken at the same time and place. Table II., p. 65, shows the extent of these variations; and the question arises as to whether the averages given in this table are taken over sufficient numbers, and therefore fairly represent the average condition of fish in each period and locality; or whether the fluctuations are due to the ordinary variations in the weight of fish of the same length.

With the view of throwing further light on this question, the average weight, per fish of standard length, has been calculated on large numbers of fish taken in different localities in each period. The calculations which are given in Table I. refer to female salmon taken on the sea coast or immediately after entering the river. It is true that Kelso, in the neighbourhood of which one lot of fish was caught, is some 16 or 17 miles above the tideway; but it would seem a fair inference that these fish have entered the river in the period in which they were caught, and that they should be treated as mouth and not as upper-water fish, since they are altogether of a larger class than those taken at Berwick-on-Tweed in July; and, since they resemble in size the large mouth fish examined by Dr. Noël Paton in October and November, being in marked contrast to the samples obtained by him from the upper waters in these months.

TABLE I.

WEIGHT PER FISH OF STANDARD LENGTH—FEMALE SALMON.

PERIOD.	LOCALITY.	Average Weight.	Average Length.	No. of Fish.	Average Weight per Fish of Standard Length.
		GRMS.	CM.		GRMS.
May and June,	Berwick-on-Tweed, -	4300	73·8	232	10723
July and Aug.,	Berwick-on-Tweed, -	6300	82·9	221	11723
Do.,	Dysart and Buckhaven (Fifeshire Coast), -	5942	80·2	20	11511
Do.,	Tayport, - - -	6531	82·6	40	11410
Do.,	N. Esk District (Sea Coast), - - -	7166	84·6	65	11836
Oct. and Nov.,	Kelso, on Tweed, - -	7711	87·76	85	11408
Do.,	Fochabers, on Spey,* -	8164	89	94	11580

The figures given in the last column show :—

1. That the difference in the average weight, per fish of standard length, in July and August on five different parts of the coast does not exceed 4 per cent.

2. That in October and November the variation in the average weight, per fish of standard length, from the lower waters of the Tweed and Spey respectively is less than 1½ per cent.

The large number of fish dealt with, the wide area over which they were taken, and the very slight variation in the results, seem collectively to afford good grounds for believing that these figures represent with a considerable degree of accuracy the condition of female salmon coming in from the sea in each of the periods specified.

A comparison of these averages with those of the estuary fish examined by Dr. Noël Paton (which are given in Table IV., p. 66), shows that the average weight per fish of standard length of the latter is in May and June 5·3 per cent. below the former, in July and August 7·3 per cent. below, and in October and November not quite 1 per cent. above. The smallness of the differences in these percentages would seem to indicate that the fish examined by Dr. Noël Paton may be taken as fair samples of fish coming in from the sea in the periods named, and, in any case, that the excess of weight, per fish of standard length, of the mouth fish as compared with the upper-water fish is not due to his samples of the former being fish in exceptionally good condition.

A similar calculation has been made with regard to males, and is given in the following table :—

---

* The fish taken at Fochabers were weighed and measured when alive. Their weight, therefore, in proportion to their length, is rather greater than that of fish which were weighed a few hours after being killed.

TABLE II.

AVERAGE WEIGHT PER FISH OF STANDARD LENGTH—MALE SALMON.

PERIOD.	LOCALITY.	Average Weight.	Average Length.	Number.	Weight per Fish of Standard Length.
		GRMS.	CM.		GRMS.
May and June,	Berwick-on-Tweed,	4784·8	76·41	43	10771
July and Aug.,	Berwick-on-Tweed,	7275·5	86·4	123	11400
Do.,	Dysart, Buckhaven, Tayport, and N. Esk Districts,	9071·8	91·9	22	11700
Oct. and Nov.,	Kelso, on Tweed,	6518·1	83·4	98	11224
Do.,	Fochabers, on Spey.*	10750	98·5	21	11248

A study of this table shows :—

1. That in July and August there is a difference of only 2¼ per cent. in the weight, per fish of standard length, of salmon taken at Berwick, and of those taken at Dysart, Buckhaven, &c.

2. That in October and November there is practically no difference in the weight, per fish of standard length, of salmon from the Spey and Tweed respectively.

A comparison of the figures given in the above table with those given in Table XIV., p. 71, shows that the average weight per fish of standard length taken at the mouths of rivers, which were examined by Dr. Noël Paton, was in May and June 18 per cent. below that of the Berwick fish, in July and August 7 per cent. below the mean of the averages of the fish taken from the different localities, and in October and November 14 per cent. below the mean of Tweed and Spey salmon. The male fish examined by Dr. Noël Paton are thus not so typical of their class as are the female fish.

Lastly, if the figures in the above table be compared with those given in Table I., it will be seen that, as regards the weight, per fish of standard length, there is a most striking similarity in males and females coming in from the sea. In May and June it is practically the same ; in July and August the excess in the weight of females over males is under 1 per cent. ; and in October and November it is barely over 2 per cent.

Material, unfortunately, is wanting to enable the same test to be applied to the samples of the upper-water fish.

* The fish taken at Fochabers were weighed when alive. Their weight, therefore, in proportion to their length, is rather greater than that of fish from other localities which were weighed a few hours after being killed.

II.—THE SOURCE FROM WHICH SALMON OBTAIN NOURISHMENT AND THE EXCHANGE OF MATERIAL IN THE BODY DURING THEIR SOJOURN IN FRESH WATER.

A. THE POWER OF THE ALIMENTARY CANAL TO DIGEST AND ABSORB FOOD.

In regard to this question the following evidence is at present forthcoming. Miescher Ruesch states *(loc. cit.)* that the stomach and gullet of the fish taken at Basel, about 500 miles up the Rhine, were contracted and folded, contrasting strongly with the distended stomach and gullet of the salmon taken in the East and North Sea. The stomach and intestine contain a clear slimy material which is never acid, while the intestine and pyloric appendages contain a slimy pus-like material full of shed epithelial cells. There was never any trace of auto-digestion. The glycerin extract of the slimy matter on the addition of dilute hydrochloric acid had sometimes a slightly dissolving action on fibrin, but so slight that it was concluded that no true digestive secretion was secreted. The gall bladder in all cases was collapsed, but the contents of the intestine are described as having a more or less deeply bile-coloured appearance.

He further states that the intestines of these fish taken from the river did not show the same tendency to early putrefaction as did the intestines of sea salmon, and this, he thinks, was due to the fact that no food being taken no organisms were introduced into the stomach. In his whole series of nearly 2000 fish, he found evidences of feeding in the stomachs of two only, both male kelts. In one the scales of some Cyprinoid fish were found, in the other an acid secretion was contained in a distended stomach, possibly indicating that digestion had been going on.

From this evidence he concludes (p. 164) that " the Rhine salmon from its ascent from the sea to its spawning, and also after this, as a rule takes no nourishment."

The investigation of the Scottish Fishery Board carried on at Berwick-on-Tweed yielded results which are hardly comparable with those of Miescher, inasmuch as the fish were here captured either while still in the sea or just after leaving it.

The question is not only one of very great interest, but it is of prime importance as regards the further investigation of the changes in muscles, ovaries, and testes.

On considering Miescher's results it seemed to us desirable to repeat some of his observations and to extend the scope of the enquiry. With this purpose Dr. Gulland has studied the microscopic changes in the digestive tract, while Dr. Gillespie has investigated the digestive activity and the bacteriology of the stomach and intestine.

3.—THE MINUTE STRUCTURE OF THE DIGESTIVE TRACT OF THE SALMON, AND THE CHANGES WHICH OCCUR IN IT IN FRESH WATER.

By G. LOVELL GULLAND, M.A., B.Sc., M.D., F.R.C.P.E.

The digestive tract of the salmon has not hitherto been the subject of any very detailed examination. Its general arrangement, of course, conforms to that usual in the group of Teleosteans to which it belongs, and it may here be divided into stomach, pyloric appendages, intestine, pancreas, and liver. All of these were examined in the series of fish under consideration.

METHOD.

The same method of preparation was used in all cases in order that the results might not be affected by any deviation in this respect. As early as possible small portions of the organs above mentioned were removed from the fish and were placed immediately in a saturated watery solution of corrosive sublimate. After 24 hours they were rapidly washed in water and then passed through a series of alcohols increasing in strength. They were embedded in paraffin, cut with the rocking microtome, and fixed to the slide by my water method. One set of sections was always stained with hæmatoxylin and eosin, whilst other stains, especially the usual anilin dyes, were employed for comparison. The sections were all mounted in balsam.

LITERATURE.

The digestive tract of the salmon has never been examined microscopically with special care in recent times, but the stomach of the nearly allied trout has been described by Valatour (1), Cajetan (2), and Oppel (3).

The most important contribution to the subject hitherto has been that by Miescher (4), who found that the stomach and gullet of the fish taken at Basel, far up the Rhine, were contracted and folded, contrasting strongly with the distended stomach and gullet of the salmon taken in the Baltic and North Sea. His results are described on p. 165 *et seq.*

It seemed desirable to re-examine the subject, and especially to subject the digestive tract of the salmon under various conditions to minute histological investigation.

THE STOMACH OF SALMON ENTERING THE RIVER.

The organ is surrounded by a serous coat, while beneath this are two layers of non-striped muscle. The external layer is thin and is disposed longitudinally; the internal layer is circular and five or six times as thick as the outer one. Between the two layers run blood vessels, lymphatics, and a nerve plexus, with an unusually large number of nerve cells, often grouped together into comparatively large ganglia. Beneath the muscular layers lies the submucosa, with many large blood vessels. The muscularis mucosæ consists of two layers, an external longitudinal and more delicate, and an internal circular rather thicker layer. The mucosa may be divided into two parts, the glandular portion and the connective tissue portion which underlies the glands and which supports them. The mass of connective tissue underlying the glands is very considerable, being often nearly as thick as the glandular layer, and is itself divided into two layers by the remarkable structure described by Oppel in the stomach of the trout, the "membrana compacta" or "stratum compactum." This layer, in whatever plane the stomach is cut, is always found as a compact hyaline band lying rather nearer the muscularis mucosæ than it does to the fundi of the glands. It is, of course, pierced by the blood vessels, etc., but I have never seen muscle strands from the muscularis mucosæ passing through it. It contains no nuclei, and no structure can be made out in it by ordinary methods. Nuclei lie upon it, however, and the fibres of the connective tissue on either side are directly continuous with it (Fig. 10), and behave to all reagents in the same way as it does. It is, therefore, certainly of connective tissue origin, and is as certainly not elastic tissue. The fibrous tissue outside this layer is more delicate than that inside it, and the inner layer contains more blood vessels, especially large venous spaces, than the outer one. A special feature of both layers is the great number of large eosinophile leucocytes to be found in the meshes of the connective tissue ; in fact, all the many leucocytes present here seem to be of this variety (Fig. 10).

In the epithelial lining of the mucous membrane the cells are of three types :—

1. The superficial epithelium and that forming the ducts of the glands.
2. The intermediate epithelium of the glands.
3. The zymin-forming epithelium of the glands.

1. The superficial epithelium (Fig. 8) is precisely like that found in most fish stomachs—long cylindrical or columnar cells with oval nuclei about the middle of the cell. Where they are best developed, on the projections between the mouths of the glands, they have a narrow clear hem on the surface, and below that a portion of the cell, which looks more hyaline and stains rather more deeply than the rest, and is sharply marked off from the deeper portion, which shows an ordinary protoplasmic structure. These cells are not goblet-cells, and no such cells are found in the stomach. The glands are very thickly set over the stomach and open by wide ducts, down which the superficial epithelium extends unaltered for some distance, usually down to the point where the glands begin to branch. Oppel notes that this branching or opening

of several secreting tubes into a common duct occurs very frequently in the trout, and I think that in the salmon it is still more frequent.

2. Before the branching of the glands begins, the characteristic upper end (Oppel's, Obereade) of the cell becomes less distinct and soon ceases to be evident, and the whole of the cell body becomes granular and undifferentiated. This I call the intermediate epithelium (Fig. 9). In the glands of the cardiac or anterior part of the stomach this part of the gland is very short, but in the pyloric region it occupies a large part of the gland.

3. The zymin-secreting epithelial cells do not exhibit any transition from the intermediate ones, but are sharply distinguished from them (Fig. 9). They are cubical, have a large rounded nucleus poor in chromatin, and a cell body with a very evident spongioplasm. The surface next the lumen of the gland is not differentiated in any way. In the cell body are scattered numerous granules which stain with eosin, though not very strongly, and which can be brought out best by M. Heidenhain's iron-hæmatoxylin, with which they stain an intense black. I have no doubt, from the analogy of the granules in the pancreas, that these are zymogen granules. They vary slightly in size, and are always most numerous towards the free end of the cell. Their actual number, or at least the ease with which they can be demonstrated, varies very much in different stomachs, and even in different parts of the same stomach, and probably would be found to depend upon whether the organ had been recently called on to digest at the time of death or not.

The amount of this variety of epithelium present in the glands varies greatly with the part of the stomach examined. About one-half or one-third of the vertical extent of the cardiac glands is occupied by this epithelium (Figs. 1, 4, 7), while at the pylorus there may be only a few cells of this sort at the bottom of a tube formed in about equal proportions of the superficial and intermediate epithelia. Perhaps one-twelfth or less of the whole extent in the pyloric glands is occupied by this epithelium. The glands intermediate in position between these two regions are intermediate also in this respect.

The above description of the normal stomach is drawn up from examination of the stomachs of seven fish caught at Berwick in March 1896. Portions of their stomachs were fixed in sublimate solution as soon as the fish were killed.

Stomachs of Salmon from Upper Waters of Rivers.

The stomachs of fish killed in the upper reaches of rivers present very striking differences from the normal type, and these seem to be caused by a desquamative catarrh of the mucous membrane. The muscular and submucous layers show no change, the muscularis mucosæ is unaltered, but the connective tissue of the mucous membrane looks swollen and hyaline. This appearance is best seen in the stratum compactum, which is thicker than in the normal stomach, and often much more folded, as though its total volume were increased. The number of eosinophile leucocytes in the meshes of the connective tissue is also usually increased. But the principal change is in the epithelium of the glands. In the extreme cases this is almost entirely desquamated, even from the fundi (Figs. 2 and 3); but usually more or less of the zymin-secreting epithelium is preserved *in situ*, though in a degenerated state. The superficial and intermediate epithelia are the first to disappear, and usually cannot be recognised at all. The fundal epithelium is never normal. The protoplasm of the cells loses its granular appearance and stains deeply and nearly uniformly with eosin, while in the nucleus the chromatin becomes massed together into a single small shrivelled deeply-staining ball, or breaks up into several smaller balls of that character. It never remains unaltered. The cells presenting these characters may either remain attached to the basement membrane of the glands or may lie in detached masses in the lumen of the glands (Fig. 2). Very often, too, leucocytes wander out from the connective tissue beneath into the glands and lie among the degenerated epithelial cells ; these are generally of the eosinophile variety, but the granules seem soon to lose their distinctive reaction, and the leucocytes can then with difficulty be distinguished from the degenerated epithelial cells.

Sometimes this mixed mass of cells lies on the surface of the mucous membrane and presents an appearance very much like that to be described in the intestine and pyloric appendages (p. 18). This material probably represents the slimy substance mentioned by Miescher.

The connective tissue septa lying between the glands fall together, and in extreme cases the appearance presented is more like that of granulation tissue from an ulcerated surface than a mucous membrane.

Now, it is of course well known that in the stomachs of the higher animals post-mortem changes, and notably post-mortem digestion, take place often with great rapidity, and it was obviously of great importance to make sure that the appearances just described were not due to some such process, especially as the stomachs of the salmon from the higher river reaches could not arrive in Edinburgh in less than several hours. In order to resolve this difficulty, Dr Noël Paton, during the course of a fishing excursion, preserved a number of trout stomachs and other parts of the alimentary tract, exactly as had been done with the salmon, some absolutely fresh, others at intervals after death of ten minutes, one hour, two hours, three hours, and six hours. These were all cut and examined, and though I shall have to say more about the pyloric appendages and intestines of these fish, it is not necessary to describe the stomachs in detail, as it was found that even six hours after death there was practically no post-mortem change. It would not have been possible to have distinguished these stomachs from those preserved at once. Even the superficial epithelium was completely preserved, and the cells of the deeper parts of the glands gave excellent demonstrations of zymogen granules when treated with iron-hæmatoxylin, while their nuclei and protoplasm remained unaltered. The connective tissue never showed the hyaline change. All these trout stomachs contained food, generally the remains of small crustaceans or insects, which they were actively digesting, and were therefore precisely in the condition where one would have expected post-mortem digestion to be specially rapid. The salmon stomachs I have described as showing this " catarrhal " change did not contain food and were therefore not active, and it seems therefore reasonable to conclude that the catarrhal change was not due to this cause. The stomach of the trout may quite fairly be compared with that of the salmon, as it is merely a miniature copy of it. The main points of difference between the two are that in the trout the glands are shorter, the stratum compactum is very much thinner, the connective tissue less in amount, and the muscular coats, of course, not so strongly developed. The minute structure of the different elements of the organ is precisely similar in the two fish.

Moreover, in some stomachs from the salmon, notably in that from No. 72, which was caught at sea, we had an opportunity of seeing the change which really takes place post-mortem, as this salmon had been out of the water 36 hours or more before it reached us. In this case the connective tissue has the usual appearance, without the hyaline change, the zymin-secreting cells are well preserved and show no chromatolysis of their nuclei, and it is only the superficial epithelium which is altered. In these cells the nuclei are shrunken, the cell-bodies are swollen and agglutinated, so that their outline has become less distinct, and they look much like the cells of mucous glands after treatment with water. But the cells retain their arrangement, and are not cast off as in the fish from the rivers.

Stomachs of Kelts.

It was exceedingly interesting to find that these had regained the normal appearance. The fish did not reach the laboratory till many hours after death, so that there was a certain amount of the post-mortem change in the superficial epithelium noted above. The zymin-secreting epithelium, however, and the rest of the mucous membrane were quite normal in every way (Fig. 4).

The Pyloric Appendages and Intestine.

The appendages in the salmon are very numerous, and some doubt has always prevailed as to their real significance, whether they are to be regarded as secreting or absorbing organs. I think there can be no doubt that, in the Salmonidæ at all events, they fulfil the latter function, for in structure they exactly resemble the upper part of the intestine, so much so in fact that but for the difference in size it would be impossible to say whether a section came from one or the other. The digestive action which their contents have been shown, in another part of this research, to possess, (p. 34) is probably due to the secretion of the pancreas. The ducts of this organ appear to open into the appendages, and its secretion would naturally impart to the contents a digestive action.

The intestine and appendages both have a peritoneal investment, an external longitudinal and internal circular and thicker muscular coat, no muscularis mucosæ, a well-marked stratum compactum exactly the same in character as that found in the stomach, a mucous membrane thrown into longitudinal folds, and covered by a cylindrical epithelium with the usual striated hem, and containing numerous chalice cells (Figs. 5, 11, 12). The only difference between the two organs is that, generally speaking, the folds in the intestine are set more closely together, so that the deepest parts of the folds look in sections like short tubular glands or Lieberkühnian follicles, whilst in the pyloric appendages the folds are more open and there is no likelihood of imagining the existence of glands at their bases. In both structures the eosinophile leucocytes are numerous, but are to be found mainly in the connective tissue about the stratum compactum. The leucocytes in the folds and passing through the epithelium are usually either of the hyaline or the smaller oxyphile variety. The epithelial cells retain their striated hem (or fringe of processes, as it probably really is), right down to the bottom of the folds.

On opening either of these structures in the fresh state there is always in the lumen a semi-fluid pultaceous mass varying in consistence between jelly and pus, and more or less yellow in colour. This is the case alike in the salmon from the river mouth, from the upper reaches, and from the sea. In the lower gut in this material are numerous more opaque hard masses. These are composed of large crystals of carbonate of lime held together by mucus. In none of the very many specimens examined was the smallest trace of undigested

B

food found in the intestinal tract. In the intestine of the trout examined the remains of food were constantly present, and appeared to consist mainly of parts of the exoskeleton and appendages of small insects and crustaceans. In one case similar material was found in the pyloric appendages, in other cases these were either empty or merely contained entozoa.

On microscopic examination of the appendages and intestine of the salmon it is found that the pus-like material is due to a desquamative catarrh exactly like that found in the stomach (Fig. 6). The mass in the tube is made up mainly of rounded cells staining deeply with eosin, and having their nucleus rounded, somewhat varying in size, but always staining deeply and uniformly with hæmatoxylin. No chromatin network can be made out. In addition a few leucocytes may be present and one or two shrunken goblet cells. The main mass of cells is certainly derived from the degeneration of the columnar epithelium, which in these cases is shed almost entirely from the folds of the mucous membrane. It is almost always easy to find some spot between the folds where this epithelium remains unaltered, and within quite a short distance one can trace every stage of degeneration until the same form as that lying free in the lumen is reached. Fig. 13 will make this clearer than any description.

Sometimes the débris in the tube is more granular and the cellular structure less easily made out, but that is evidently merely a later stage of the same process.

I have never found a salmon in the intestine and pyloric appendages of which (for the two are always at the same stage of the process, another proof of their identity of function) this change was not present to a greater or less extent. The pyloric appendages and intestine of the seven Berwick salmon were examined, and as these had all been fixed immediately after death with sublimate, there could be no question of post-mortem change. In four out of the seven the change was practically complete, either the entire epithelium was degenerated, or only a little was left deep down between the folds. In the remaining three the change, though present, was less marked, being confined mainly to the apices of the folds; in one there was very little catarrhal change. In all the fish from the higher reaches, from the sea, and in the kelts from the river the change was complete. It is very curious to see the connective tissue framework of the folds, with its blood vessels much congested as a rule, lying absolutely bare of epithelium in this pus-like mass (Fig. 6).

This change must certainly be a pathological, catarrhal, or seasonal one, for in the trout we see nothing of it. There the columnar epithelium everywhere covers the folds, both in the specimens fixed immediately after death, and in those fixed at varying times post-mortem, except in the intestine of the specimen fixed six hours after death. Here there is indeed a considerable desquamation of epithelium, which is lying free in the lumen of the gut, but the cells are not degenerated in any way. They still exhibit the characteristic hem, their nuclei show the chromatin network, and the desquamation is probably accidental, and due to some squeezing of the specimen before it was fixed.

There is one source of fallacy in these cases, or rather what might be regarded as a source of fallacy, that is the presence of parasitic worms in the intestine and pyloric appendages of the salmon. There are few fish which are entirely free from these, and sometimes the intestine is greatly distended with them. It might be thought that these were the cause of the catarrh, and indeed one finds on section that in some of the intestines which contain them there is nothing left but the muscular coats, and a mere shred of connective tissue infiltrated with leucocytes. But they are almost as common in the trout as in the salmon, and there their presence does not produce a desquamative catarrh, but simply a flattening out of the folds of the mucous membrane from distension ; the epithelium remains intact. It is noticeable, however, that when these worms are present in the trout the number of goblet-cells in the epithelium is always unusually large, sometimes, indeed, the goblet-cells considerably exceed the ordinary epithelial cells in number, as if the irritation caused by the presence of the worm necessitated a greater flow of mucus to lubricate the surface of the mucous membrane.

THE PANCREAS.

This is not gathered into a compact organ, but the acini are scattered more or less diffusely through the long strands of intraperitoneal fat lying round the stomach, pyloric appendages, and intestine (Fig. 14). The microscopic ducts—one can hardly call them interlobular or intralobular, as there are no true lobules—are lined with a cylindrical epithelium, and resemble the ducts of mammalian salivary glands much more than those of the mammalian pancreas. The cells of the secreting acini are like pancreas-cells elsewhere, and they evidently go through exactly the same cycle of appearance during the formation of and extrusion of zymogen granules as those described by Langley. One point of difference there is, however, inasmuch as in the trout and salmon the zymogen granules in the pancreas, as in the stomach cells, stain a deep black with iron-hæmatoxylin, which renders it very easy to make out the stage of secretion. Fig. 15 shows a pancreatic acinus where the cells are full of granules, the so-called stage of "rest," when no extrusion of granules is going on, but where they are being heaped up in the cell ready to aid the process of digestion. Fig. 16 depicts an acinus where the granules have mostly been extruded, the so-called "active" stage. The only pancreas where I found the cells completely full of granules was that of a trout which had been preserved immediately after death. All the other trout and all the Berwick salmon showed the cells more or less emptied of granules. The pancreas was not often present in the portions of salmon from the rivers fixed for examination, but where it was to be seen the cells were generally shrunken and shrivelled, and contained no granules.

THE LIVER.

Comparatively few livers were examined, and these showed no very marked difference beyond the presence or absence of fat. This is brought out better, however, by chemical analysis (p. 100) and need not here be discussed. Generally speaking, however, it was found that the livers of the salmon taken at the river mouths contained much fat (Fig. 18), while those of the salmon from the upper reaches, and the kelts, were deficient in it (Fig. 17). The fat, where present, was generally in the form of large droplets in the cells, and was distributed pretty equally throughout the organ.

THE GALL BLADDER.

The following are the statistics with regard to the state of the gall bladder :—

(a) Fish at the Mouth of the River;

	Distended.	Empty.
May - - -	9	3
June - - -	4	0
July - - -	10	2
August - -	3	2
October - -	2	4
November - -	0	7

or, in Percentage of Fish Taken.

	Distended.	Empty.
May - - -	75	25
June - - -	100	0
July - - -	83	17
August - -	60	40
October - -	33	67
November - -	0	100

(b) Fish in Upper Waters.

In all the gall bladders were collapsed.

(c) Kelts.

In most of these the gall bladders were distended.

I examined several gall bladders microscopically, and they all, whether distended or empty, whether from fish at the mouth of the river, in the upper waters, or from kelts, showed a desquamative catarrh of exactly the same kind as that described in the stomach, intestine, etc. In some the epithelium had entirely disappeared, while in others part of it could be seen in a degenerated state, either lying detached in the lumen of the bladder, as was more usual, or attached in fragments to the wall. There is no stratum compactum in the gall bladder; the connective tissue was in a state of more or less evident hyaline degeneration.

CONCLUSIONS.

We may take it for granted that at some period of its existence the salmon's alimentary tract has the same normal structure as that of the trout, and it is evident that its sojourn in the sea is the time of normal digestive activity. Probably for some time before the fish enter the river, and certainly while they are lying at the mouth of it, the catarrhal change begins, and begins clearly in the intestine and pyloric appendages; the stomach is at that time unaffected. By the time the fish have reached the upper waters the stomach has been attacked, and the whole digestive tract is in a state of catarrh. After spawning is over, the stomach is the first part to recover, and in the kelts it is again histologically normal, while the intestine and pyloric appendages probably recover when the fish have returned to the sea.

It is evident that this desquamative catarrh is not caused by the action of fresh water either on the general health of the fish or locally on the parts of the alimentary canal, for in many fish taken from the sea the change was already complete in the intestine and appendages.

Finally, it is well again to emphasise the fact that in no part of the alimentary canal of the many fish examined, including kelts, were any remains of undigested food discovered upon microscopic examination.

LITERATURE.

1. *Valatour M.* Recherches sur les glandes gastriques et les tuniques musculaires du tube digestif dans les Poissons osseux et les Batraciens. Ann. des Sc. nat. 4. Sér. Zool. T. 16. 1861.

2. *Cajetan, J.* Ein Beitrag zur Lehre von der Anatomie und Physiologie des Tractus intestinalis der Fische. Inaug.-Diss. Bonn. 1883.

3. *Oppel, A.* Lehrbuch der vergleichenden mikroskopischen Anatomie. 1er. Teil. Der Magen. Jena. 1896.

4. *Miescher.* Fischerei-Ausstellung zu Berlin. 1880.

DESCRIPTION OF THE FIGURES.

The following letters are used to indicate the same structures throughout:—

ser.,			serous coat.	eos. l.,	eosinophile leucocyte.
m. long.,	longitudinal muscular coat.			muc.,	mucosa.
m. circ.,	circular muscular coat.			s. muc.,	submucosa.
m. m.,	muscularis mucosæ.			ep. m.,	epithelial part of mucous membrane.
c.t.,	connective tissue.			s. ep.,	superficial epithelium.
c.t.f.,	connective tissue fibre.			int. ep.,	intermediate epithelium.
c.t.n.,	connective tissue nucleus.			z. ep.,	zymin-secreting epithelium.
s.c.,	stratum compactum.			z. gr.,	zymogen granules.
b.-v.,	blood-vessel.			col. ep.,	columnar epithelium.
art.,	artery.			ch. c.,	chalice cell.
ven.,	vein.			deg. ep.,	degenerated epithelium.
cap.,	capillary.			p.d.,	pancreatic duct.
r.b.c.,	red blood corpuscle.			ft.,	fat.
l.,	leucocyte.				

Figs. 1 to 6 are micro-photographs of actual sections; the lens used was Leitz 3 and ocular 4. The remaining Figures are drawings by the author. The outlines were, in all cases, filled in with Zeiss's camera lucida. For Figs. 8, 9, 10, 15, 16, 17, 18, the objective was Zeiss's homog. immers., apochr. 2·0 mm., apert. 1·40 with compensat. ocular 8: and the magnification is therefore × 1000; for Figs. 12 and 13, the same objective, with compensat. ocular 4, magnification × 500; for Fig. 7, Zeiss DD., with ocular 2, magnification about × 200; for Figs. 11 and 14, Leitz 3 with ocular 4, magnification about × 60.

Fig. 1. Stomach of salmon from mouth of river (Berwick), showing normal mucous membrane. Iron-hæmatoxylin.

Fig. 2. Stomach of salmon from upper waters, showing early stage of degeneration and desquamation, and hyaline change in connective tissue. Iron-hæmatoxylin.

Fig. 3. Stomach of salmon from upper waters, showing very advanced stage of degeneration of the mucous membrane. Iron-hæmatoxylin.

Fig. 4. Stomach of kelt, showing regenerated mucous membrane. The secreting epithelium of the glands is loaded with zymogen granules, which are stained black. Iron-hæmatoxylin.

Fig. 5. T.S. Intestine of salmon from the mouth of river (the only one where there was no catarrhal change), showing the folds of the

22   *Investigations on the Life-History*

mucous membrane cut across and covered with columnar epithelium. Hæmatoxylin and eosin.

Fig. 6. T.S. Intestine of salmon from upper waters, with catarrhal change far advanced. Only a few patches of epithelium at the bottom of the folds remain attached, and these are degenerated. Hæmatoxylin and eosin.

Fig. 7. Stomach of salmon from mouth of river, to show normal arrangement. Hæmatoxylin and eosin.

Fig. 8. Superficial epithelium of stomach (estuary salmon). Iron-hæmatoxylin.

Fig. 9. Junction of intermediate and zymin-secreting epithelium from gastric gland (estuary salmon). Iron-hæmatoxylin.

Fig. 10. Stratum compactum from gastric mucous membrane (estuary salmon), to show its continuity with the neighbouring connective tissue-fibres, and the number of eosinophile leucocytes about it. (The details of the nuclei of the leucocytes have been omitted). Hæmatoxylin and eosin.

Fig. 11. T.S. through whole thickness of wall of pyloric appendage (estuary salmon), to show arrangement and the identity of its structure with that of the intestine. Hæmatoxylin and eosin.

Fig. 12. Portion of a longitudinal section through a fold of intestinal mucous membrane (estuary salmon). Hæmatoxylin and eosin.

Fig. 13. Intestine of salmon from mouth of river, showing a patch of normal epithelium at the bottom of one of the folds, and the process of degeneration and desquamation which the cells are undergoing. Hæmatoxylin and eosin.

Fig. 14. Pancreas (estuary salmon), showing its distribution in the fat lying between the pyloric appendages. Hæmatoxylin and eosin.

Fig. 15. Pancreas (trout), showing an acinus in the "resting" stage, with the cells full of zymogen granules. Iron-hæmatoxylin.

Fig. 16. Pancreas (trout), showing an acinus in the "active" stage, when most of the granules have been discharged. Iron-hæmatoxylin.

Fig. 17. Liver (salmon), from upper water, with only a small amount of fat in the cells. Hæmatoxylin and eosin.

Fig. 18. Liver (salmon), from sea, the cells distended with fat globules. Hæmatoxylin and eosin.

PLATE I.

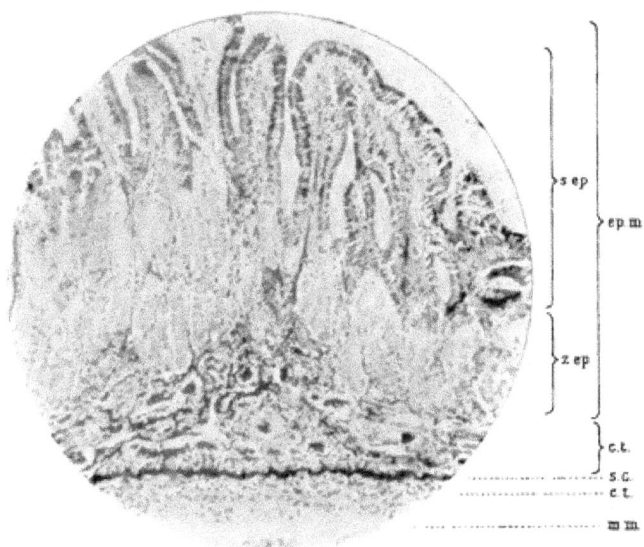

FIG. 1.

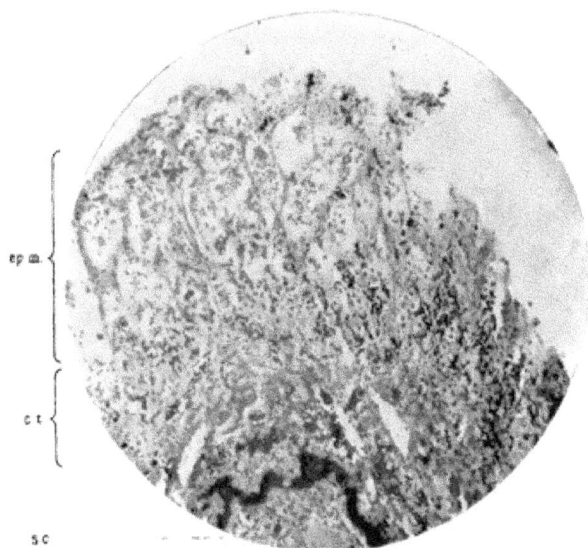

FIG. 2.

PLATE II.

muc

m m

s muc.

m circ

FIG. 3.

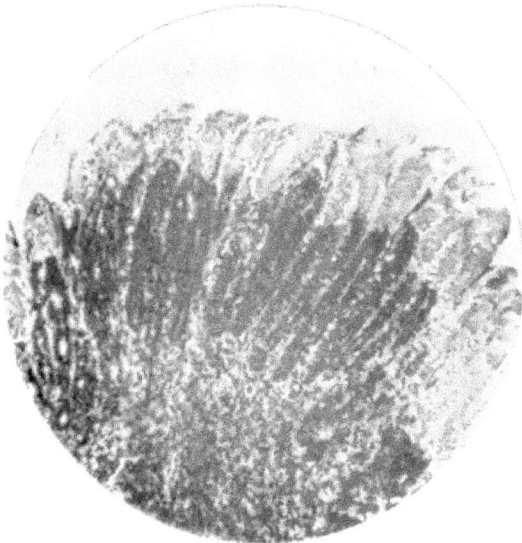

s ep

z ep

ct sc

FIG. 4.

PLATE III.

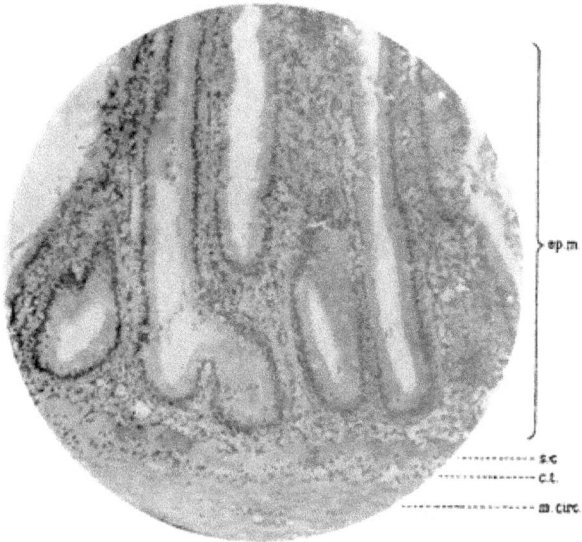

ep.m.

s.c.
c.t.
m. circ.

FIG. 5.

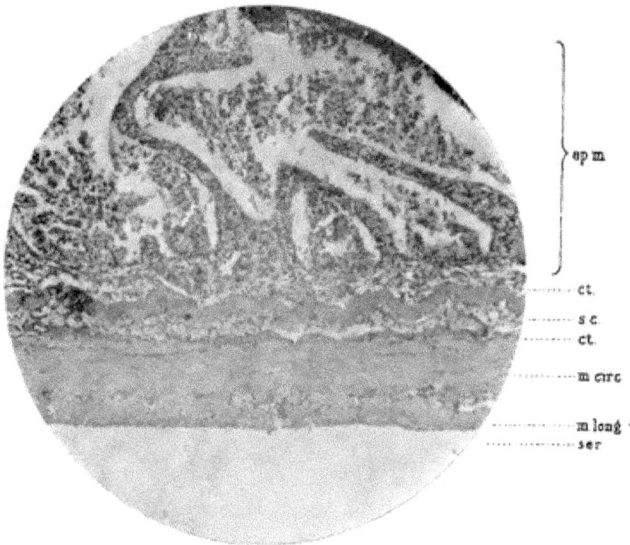

ep.m.

ct.
s.c.
ct.
m. circ.
m. long.
ser.

FIG. 6.

PLATE IV.

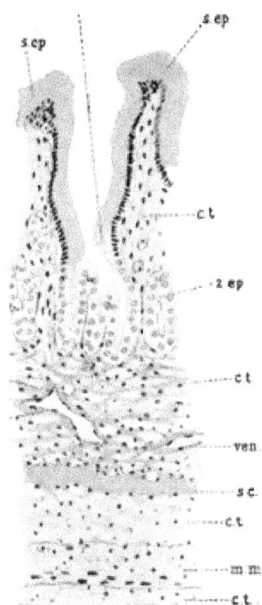

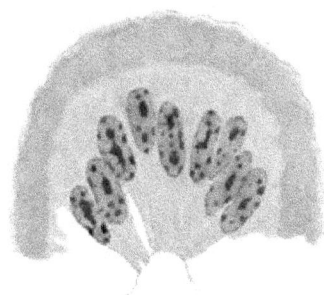

Fig. 7.

Fig. 8.

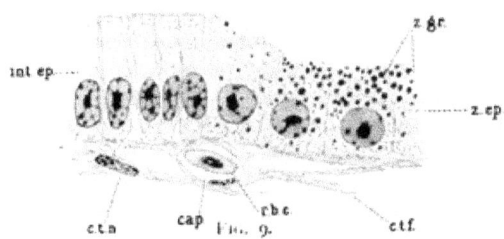

Fig. 9.

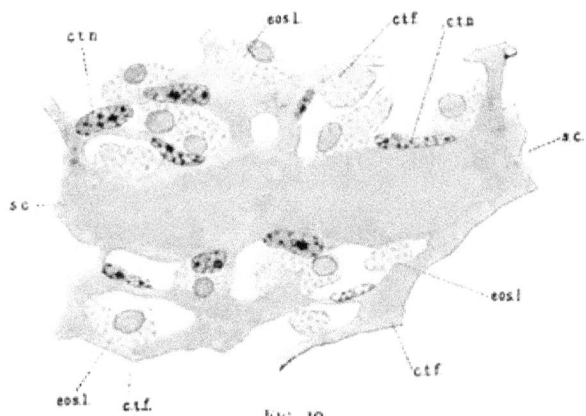

Fig. 10.

PLATE V.

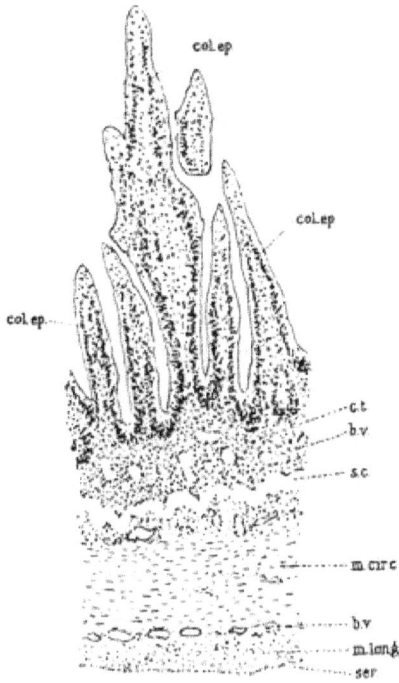

FIG. 11.

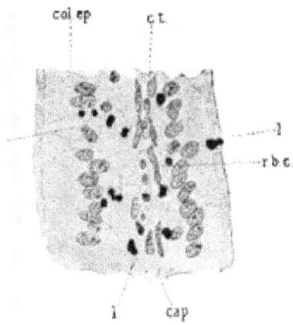

FIG. 12.

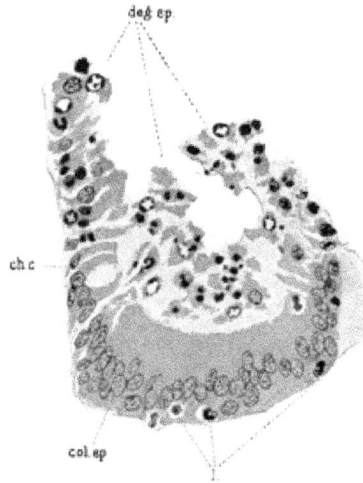

FIG. 13.

PLATE VI.

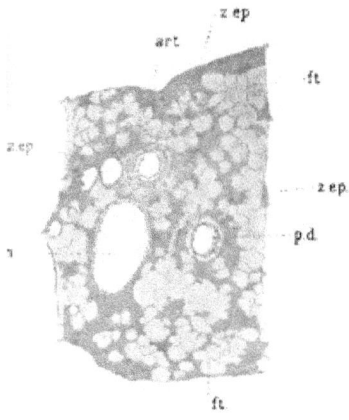

FIG. 14.

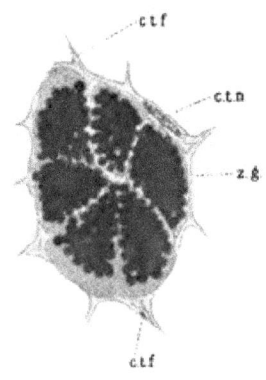

FIG. 15.

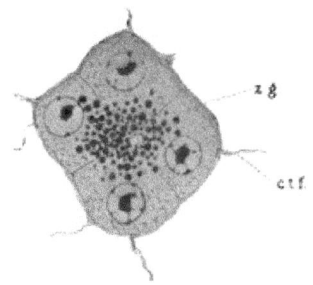

FIG. 16.

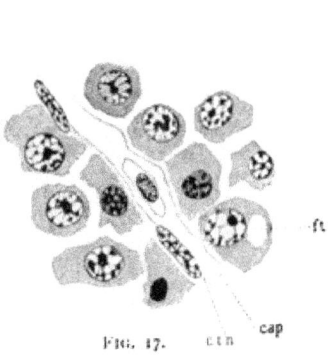

FIG. 17.

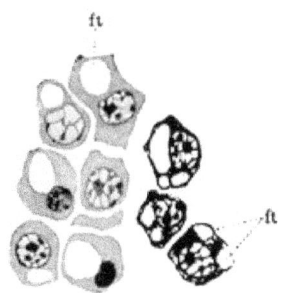

FIG. 18.

4.—CHANGES IN THE DIGESTIVE ACTIVITY OF THE SECRETIONS OF THE ALIMENTARY CANAL OF THE SALMON IN DIFFERENT CONDITIONS.

By A. LOCKHART GILLESPIE, M.D., F.R.C.P.E., F.R.S.E.

1. PRELIMINARY.

As already pointed out, the question of whether salmon feed during their sojourn in the river has been much debated. The absence of food from the alimentary canal has been variously explained by activity of digestion or by the power of rejecting the contents of the stomach when the fish is captured.

Dr. Gulland's investigations have demonstrated the occurrence of a degeneration of the mucous membrane of the alimentary tract during the stay of the fish in the river, and very clearly indicate that little or no absorption of food can go on.

In this investigation the digestive activity of the various parts of the alimentary tract is considered.

The points investigated are the proteolytic and diastatic powers of extracts made from the mucous membrane lining the stomach, the bowel, and the pyloric appendages, and the variations in these in relation to the season and the part of the river from which the fish were obtained.

It is well known that the peptic digestion of proteids in cold-blooded animals owes its activity at low temperatures—it can take place at 0° C. —to the large proportion of pepsine present in the secretion of the gastric glands.

There are two considerations concerning the degree of activity of enzymes, or unformed ferments, which apply to the subject under discussion. The first is that the amount of digestive action exerted by any agent is in direct proportion to the quantity of enzyme present, while the second is that enzymes are more active in media as the temperature rises from 0 deg. to 40 degs. Cent., their power gradually diminishing as the temperature rises above 40 degs., until at 70 degs. all activity ceases.

In fish the large proportion of pepsine allows rapid proteolysis to take place in the stomach, although the temperature at which it occurs may be low, and if the active secretion be tested outside the body at a higher temperature, a very much more powerful action is exerted on proteids than by the ordinary peptic extract obtained from the gastric mucous membrane of mammals.

Charles Richet, in his work entitled *Du Suc Gastrique chez l'Homme et les Animaux*, proved that the gastric juice in marine fish, such as the dog-fish, is almost neutral when fasting, and mucoid in character. During digestion the secretion is much more acid than in man, and contains a more active pepsin. In a dog-fish weighing 1 kilogramme he obtained 5 grammes of dried gastric mucous membrane, which was capable of digesting 150 grammes of egg albumin, or nearly one-sixth of the total weight of the fish. In Scyllium he found the gastric acidity to be as high as 0·79 per cent. HCl.

Krukenberg (*Untersuchungen aus dem phys. Instit. der Univ. Heidelberg*, 1882, II., p. 396) states that the pepsine of the stomach of fish, or an analogous body to pepsine, acts as well at 20°C. as at 40°C., and is most powerful in solutions of hydrochloric acid containing 1, 1·5, or 2 grammes per litre, but is active in solutions with 10, 15, or 20 grammes of this acid in the litre.

The frequent absence of the remains of food in the stomach of feeding fish very shortly after its ingestion led some to the idea that the rapidity of digestion might be accounted for by the action of the bacteria in the stomach in assisting the gastric ferment, or might even be due to the unassisted action of the bacteria.

Richet, however, has showed (*loc. cit.*) that an extract of the gastric mucous membrane of fish is capable of digesting albumin in a 2·3 per cent. solution of hydrochloric acid, and in the presence of ether or chloroform in excess, or of cyanide of potash. Under none of these conditions would bacteria be able to exist.

In herbivorous fish, such as the carp and tench, a diastatic ferment is present in the gastric secretion. In carnivorous fish this ferment is absent.

In 1880 Miescher Ruesch published an important paper founded on his observations on the physiology of salmon caught in the Rhine during the preceding eight years. He agreed with Barfurth, His, and Glaser, that "the Rhine salmon takes no food from the time when it leaves the sea until it has spawned, and seldom even after this," until it again reaches the salt water. No remains of food were found by him in any of the winter and spring fish examined. The mucus occurring in the stomach was never acid in reaction. No active digestive ferment seemed to be secreted, while a glycerine extract of the gastric mucous membrane exhibited little proteolytic power.

In a male fish, however, coming from the spawning beds—a kelt—he found in a large, flabby stomach two fairly large fish whose anterior portions were already digested. In another male kelt, although no trace of food was present, he found that the thin secretion of the stomach was of acid reaction.

To sum up the outcome of Miescher Ruesch's work, it would seem that salmon ascending streams to the spawning grounds do not feed, and that if they swallow portions of food they are unable to digest them. After they have spawned, the males occasionally take food on their way down to the sea; the female fish may or may not do so, but Miescher Ruesch failed to find any evidence in support of this, while as regards the male fish, the gastric glands have become so far more active that their secretion possesses an acid reaction.

2. THE MATERIAL EXAMINED.

In Table I. the details regarding the date of capture, the part of the river from which the fish were obtained, the number of fish used for investigating the peptic and tryptic power of their extracts, and the total number of salmon employed are arranged in tabular form :—

[TABLE.

TABLE I.

TABLE OF FISH EMPLOYED.

For Digestive Activity of Gastric Extract.

No.	Year.	Month.	No. of Fish.	Part of River.		Kelts.
				Mouth.	Upper.	
1	1895.	July.	X.	–	1	
2	,,	,,	XI.	1	–	–
3	,,	,,	XII.	1	–	–
4	,,	,,	XIII.	1	–	–
5	,,	,,	XIV.	–	1	1
6	,,	August.	XIX.	1	–	–
7	,,	September.	XXIV.	1	–	–
8	,,	October.	XXIX.	1	–	–
9	,,	,,	XXX.	1	–	–
10	,,	,,	XXXI.	1	–	–
11	,,	November.	XXXVII.	–	1	–
12	,,	,,	XXXVIII.	–	1	–
13	,,	December.	XL.	1	–	–
14	1896.	March.	I.	1	–	–
15	,,	,,	II.	1	–	–
16	,,	,,	III.	1	–	–
17	,,	,,	IV.	1	–	–
18	,,	,,	V.	1	–	–
19	,,	,,	VI.	1	–	–
20	,,	,,	VII.	1		
21	,,	,,	VIII.	–	1	1
22	,,	,,	IX.	–	1	1
23	,,	May.	X.	–	1	1
24	,,	,,	XI.	–	1	–
25	,,	,,	XII.	–	1	–
26	,,	,,	XIII.	1	–	–
27	,,	,,	XIV.	1	–	–
28	,,	,,	XV.	1	"	–
29	,,	,,	XVI.	1	–	–
30	,,	,,	XVII.	1	–	–
31	,,	,,	XVIII.	1	–	–
32	,,	,,	XIX.	1	–	–
			Total, ...	23	9	4

For Tryptic Digestion of Intestinal and Appendical Extract.

No.	Year.	Month.	No. of Fish.	Part of River.		Kelts.
				Mouth.	Upper.	
1.	1895.	July.	XI.	1	–	–
2	,,	,,	XIV.	–	1	1
3	,,	December.	XXXIX.	1	–	–
4	,,	,,	XL.	1	–	–
5	,,	,,	XII.	1	–	–
6	1896.	May.	XIII.	1	–	–
7	,,	,,	XIV.	1	–	–
8	,,	,,	XV.	1	–	–
9	,,	,,	XVI.	1	–	–
10	,,	,,	XVII.	1	–	–
11	,,	,,	XVIII.	1	–	–
12	,,	,,	XIX.	1	–	–
			Total, ...	11	1	1

Table 11. represents the same data arranged according to season :—

TABLE 11.

TABLE OF FISH WITH REGARD TO SEASON AND PART OF RIVER.

Seasons.	Digestive Activity.						Bacteriological.					
	1885.	1896.	Total.	Month.	Upper Water.	Kelts.	1885.	1896.	Total.	Month.	Upper Water.	Kelts.
1. March,	0	9	9	7	2	2	0	1	1	1	0	0
2. May,	0	10	10	7	5	1	0	2	2	2	1	0
June,	0	0	0	0	0	0	0	4	4	2	2	0
Total,	0	10	10	7	5	1	0	7	7	4	3	0
3. July,	5	0	5	3	2	1	5	5	10	6	4	1
August,	1	0	1	1	0	0	3	0	3	3	0	0
Total,	6	0	6	4	2	1	8	5	13	9	4	1
4. September,	1	0	1	1	0	0	1	0	1	1	0	0
October,	3	0	3	3	0	0	3	3	6	6	2	0
November,	24	0	24	24	2	0	4	3	7	5	2	0
December,	4	0	4	4	0	0	0	0	0	0	0	0
Total,	7	0	7	5	2	0	8	6	14	9	5	0
Grand Totals,	13	19	32	23	9	4	16	25	41	29	22	1

3. Methods.

The stomach and intestine of each fish used for the purpose of testing the digestive activity were placed, after removal, in methylated spirits for forty-eight hours to extract the excess of fatty matter, and to harden the tissues. They were then taken out of the spirit, cut up into small pieces, and placed in glycerine, to which a small quantity of water had been added. After a month to six weeks the glycerine extract was filtered off, and made up in each case to 50 cc. The stomachs and intestines were treated separately.

The activity of the extracts was then determined in the usual way. 5 cc. of the extracts were added to 10 cc. of an egg-albumin solution of known strength, hydrochloric acid or sodium carbonate mixed with it, and the whole made up to 50 cc. by the addition of water. The amount of hydrochloric acid added represented in each case the quantity necessary to form an acidity of 0·1095 per cent. in the 50 cc., and the sodium carbonate represented 1·5 per cent. in alkalinity.

Of the solutions 5 cc. were used to estimate the acidity or alkalinity present, and the remaining 45 cc. were left at the ordinary temperature of the laboratory for six to eight hours.

The solutions were then neutralised, boiled, and then acidified with acetic acid, a few drops of a saturated solution of acetate of soda being also added. The precipitate which formed was caught on a weighed filter-paper, washed with boiling water, alcohol, and ether, and then dried and weighed. The increase in the weight of the paper represented the amount of albumin still undigested. This subtracted from the original weight of the albumin employed gave the quantity digested. Corrections were made in these figures for the amount of coagulable proteid contained in the glycerine extract on several occasions at the commencement of the research, but as the additional precipitate was found to be so small, and to be practically constant, it has been ignored throughout.

The diastatic power of the glycerine extracts was estimated by its action on starch, a solution of iodine and iodide of potassium being employed as the indicator. A 1 per cent. starch solution was made faintly alkaline with sodium carbonate, 5 ccm. of the extract added, and the mixture left for two hours at the temperature of the room.

No action was obtained from the gastric extracts, and the further details of the experiments with these extracts are omitted. All the intestinal and appendicical extracts proved to be fairly active.

4. Digestion in the Stomach.

(a) *The Peptic Activity of the Glycerine Extracts of the Gastric Mucous Membrane.*

In Table III. the details of this part of the research are summarised :—

[Table.

TABLE III.

Peptic Power.

1. *From Mouth of Rivers.*

No.	Number of Fish.	Season.	Digestive Power. % Digested.	Acidity of Extract.
				%
1	XV.	May, 1895	33·3	0·21
2	VII.	March, 1896	33·0	0·65
3	XVI.	May, 1896	22·2	0·13
4	XVII.	,,	21·1	0·12
5	XXIX.	October, 1895	⎰ 18·3	⎰ 0·24
6	XXX.	,,	18·3	0·24
7	XXXI.	,,	⎱ 18·3	⎱ 0·24
8	VI.	March, 1896	12·9	0·61
9	XIII.	May, 1896	10·4	0·13
10	IV.	March, 1896	9·5	0·24
11	XI.	December, 1895	7·9	0·24
12	XVIII.	May, 1896	5·5	9·01
13	I.	March, 1896	5·2	0·19
14	V.	,,	2·6	0·41
15	XIX.	May, 1896	1·1	0·004
16	XL.	July, 1895	0·0	0·0
17	XII.	,,	0·0	0·3
18	XIII.	,,	0·0	0·006
19	II.	March, 1896	0·0	0·01
20	III.	,,	0·0	0·13
21	XIV.	May, 1896	0·0	0·0
22	XIX.	August, 1895	0·0	0·022
23	XXIV.	September, 1896	0·0	0·022
Total 23			9·5 %	0·167
Total with Digestive Power, 15			14·6 %	0·244

2. *From Upper Waters.*

No.	Number of Fish.	Season.	Digestive Power. % Digested.	Acidity of Extract.
				%
1	XXXVII.	November, 1895	⎰ 14·7	⎰ 0·17
2	XXXVIII.	,,	⎱ 14·7	⎱ 0·17
3	XL.	May, 1896	10·4	0·004
4	XII.	,,	6·08	0·005
5	X.	July, 1895	0·0	0·01
Total 5			9·17 %	**0·071**
Total with **Digestive Power**, 4			11·8 %	0·174

3. *Kelts.*

No.	Number of Fish.	Season.	Digestive Power. % Digested.	Acidity of Extract.
				%
1	VIII.	March, 1896	⎰ 33·9	⎰ 0·47
2	IX.	,,	⎱ 33·9	⎱ 0·47
3	X.	May, 1896	21·7	0·15
4	XIV.	July, 1895	0·0	0·009
Total 4			**22·7 %**	**0·274**

1. FISH FROM THE ESTUARIES.—Of the 23 fish caught at the mouth of the rivers the glycerine extract of the gastric mucous membrane possessed some peptic power in 15. The average percentage of albumin digested in these is 14·6, while check experiments with ordinary commercial pepsine, at the ordinary temperature and with the same proportion of hydrochloric acid, show a digestive power reaching to 72 per cent. Including all the fish caught at the mouth the digestive activity only equals 9·5 per cent.

2. FISH FROM THE UPPER WATERS.—Four of the five fish from the upper waters were found to give positive results in this part of the research, with 11·8 per cent. of albumin digested. The total number —five—give a percentage of 9·17 per cent.

These results seem to indicate that the mucous membrane of the stomach in the fish which have reached the upper waters of the rivers possesses a somewhat smaller peptic power than that of salmon caught close to the sea.

3. KELTS.—The figures obtained from the four kelts show a higher digestive activity than in the fish included in the first and second classes—namely, 22·7 per cent. One of the kelts, however, possessed no proteolytic ferment in its stomach; the remaining three possessed a digestive power of 29·8 per cent. The kelt without digestive power was one which had not returned to the sea in July.

(b) *Relation of Peptic Activity to Acidity.*

A comparison of the percentage acidities of the extracts with their proteolytic power shows a direct relationship between these, although in several cases the relationship is departed from. In the fish caught at the river mouths those in which an active digestion of albumin was found, to a greater or less extent, the mean acidity of the extracts of the stomachs reached 0·244 per cent. in terms of hydrochloric acid, against 0·023 per cent., or ten times as little, in those in which no digestive action occurred. Similarly, of the five fish from the upper waters the four presenting actual evidence of digestive power give a mean acidity in their extracts of 0·074 per cent., the remaining one a mean of only 0·01 per cent.

In Table IV. the figures relating to the comparison between the amount of albumin digested and the acidity of the extracts are given for the total number of fish used:—

TABLE IV.

RELATION BETWEEN THE DIGESTIVE POWER AND THE ACIDITY.

Totals.

	No.	Season.	Digestive Power.	Acidity.
			%	%
Total, ...	32	Whole Year.	11·1	0·166
Those with Digestive Power, ...	22	—	16·13	0·231
With no Digestion, ...	10	—	—	0·0209

In the total number, 32, the digestive power averages 11·1 per cent., the acidity 0·166 per cent. In the 22 with some actual digestive power this reaches 16·13 per cent., the acidity 0·231 per cent., while in the 10 fish in which no peptic digestion occurred the mean acidity is only 0·0209 per cent. In the first two the proportion between the amount of albumin digested and the acidity of the extract is exactly the same. The acidity per cent. is 1·45 per cent. of the amount of albumin acted on.

It would thus seem that the amount of active ferment extracted from the gastric mucous membrane appears to be exactly proportional to the acidity of the extract in the salmon examined in this research.

Table V, shows the figures relating to the peptic power and acidity of the stomach extracts arranged as to season:—

TABLE V.

FIGURES RELATING TO THE DIGESTIVE POWER AND THE ACIDITY, ARRANGED AS TO SEASONS.

Season.	Number.	Average Digestive Power.	Average Acidity.
		%	%
March, ...	9	14·55	0·33
May and June, ...	10	13·1	0·076
July and August, ...	6	0·0	0·008
September to December, ...	7	13·17	0·19

The result is very striking. The fish examined in March give a comparatively high digestive power with the highest acidity, those captured in May-June and September-December present a lower digestive power, identical in amount in both periods, though the acidity present in the fish of the latter period is almost three times that of those in former. The fish caught in July and August have no digestive power and a very low acidity.

(c) *Relation of Peptic Activity to the Micro-Organisms present.*

Table VI. is formed of figures relating to the fish in which both the digestive power of the stomach mucous membrane and the number of bacteria present were investigated. The bacteriological data will be considered more fully later on, but a direct comparison between the results of the two investigations is useful in this place :—

[TABLE.

PL

Comparison between the Acidity per cent. of the Gastric Extracts, the Peptic
Activity and the average number of Organisms grown from the Salmon.

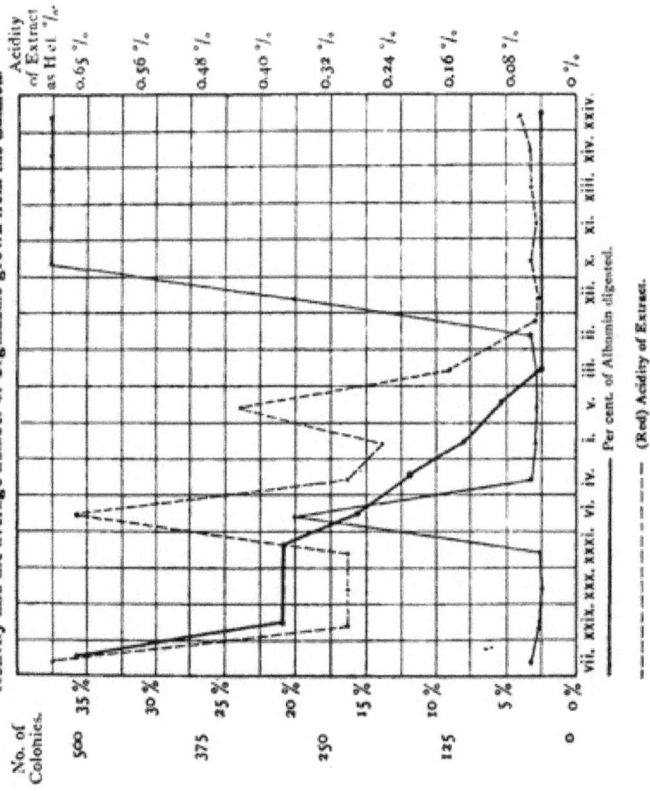

No. of
Colonies.

Acidity
of Extract
as Hcl °/₀

——————— Per cent. of Albumin digested.

———————— (Red) Acidity of Extract.

TABLE VI.

COMPARISON BETWEEN THE DIGESTIVE POWER AND THE ACIDITY OF THE
GASTRIC EXTRACTS, ARRANGED WITH REGARD TO THE MICRO-
ORGANISMS GROWN FROM THE STOMACH.

No.	Number of Colonies.		No. of Fish.	Date.	Peptic Activity.	Acidity of Extract.
	Tube I.	Tube II.				
1	0	0	XXIX.	Oct. 1895	18·3	0·24
2	0	0	XXX.	,,	18·3	0·24
3	0	0	XXXI.	,,	18·3	0·24
4	1 Mould	0	V.	Mar. 1896	2·6	0·41
5	1 Mould	0	III.	,,	0·0	0·13
6	Several Moulds	0	I.	,,	5·2	0·19
7	5, 3 Moulds, 2 Liquefying	7 Moulds	VII.	,,	33·0	0·65
8	8, 5 Moulds, 3 Liquefying	1 Mould	IV.	,,	9·5	0·25
9	12 Liquefying	1 Liquefying	II.	,,	0·0	0·01
10	Large Number Liquefying	4 Moulds, 1 Liquefying	VI.	,,	12·9	0·61
11	Large Number Pure Growth B. Coli	0	XII.	July 1895	0·0	0·0
12	Innumerable	200 (100 Coli)	X.	,,	0·0	0·01
13	,,	53 ; 3 Coli	XI.	,,	0·0	0·0
14	,,	45 ; 3 Coli	XIII.	,,	0·0	0·006
15	,,	Large Number	XIV.	,,	0·0	0·000
16	,,	,,	XXIV.	Aug. 1895	0·0	0·022
Total, 16 ...	3 None / 3 Moulds only / 3 Small Number / 7 Large Number		March 7 / July 5 / August 1 / October 3		7·4	0·187
Of the 3 with no Growth,					18·3	0·24
,, 3 with Moulds only,					2·6	0·24
,, 3 with a few Growths,					14·1	0·3
,, 7 with a Large Number					1·84	0·094
,, 1 no Liquefying Growth,					0·0	0·0
,, 9 with Liquefying Growths,					6·26	0·173

In Chart I. the relationship between the acidity, peptic activity, and bacterial contents of the stomach is graphically shown.

The first three fish noted yielded no bacteria of any kind from the alimentary tract, while their peptic activity, which was tested in common, was high, and the acidity of their common extract above the average. Three fish yielded growths of moulds only, and presented a much lower peptic power, but a similar acidity. Three also gave evidence of the presence of moulds, along with a small number of true bacteria. Their digestive power was above the average but less than in the first three, while the acidity in their extracts reached 0·3 per cent. On the other hand, in the seven fish in which no digestive action was observed, the bacteria were very numerous, and the acidity low. The latter set were almost all captured during the months of July and August.

These results appear to show that there is a direct connection between the digestive activity and acidity of the stomach extract and the number and form of organisms present. When the extract is acid and of considerable digestive power organisms are either absent, or few in number. When only a slight digestive power with a similar acidity occurs, no organisms are present save moulds. Here the growth of other organisms may be arrested by the growth of the moulds

iffort>5nt>5 in the contents. A low acidity and practically an extinct
power of peptic digestion coincides with the presence of large numbers
of micro-organisms. The fish from the stomachs of which no organisms
were cultivated were all caught in October at the river mouth, and
possessed a marked power of digestion and a high degree of acidity.
The three in which moulds alone were found also came from the
estuaries during March. The three with a few bacteria in addition to
moulds were caught at the same time of the year, but gave a higher
digestive activity. Of the seven in which innumerable bacteria were
found one was caught at the mouth in March, the others in July
and August, two in the upper waters, five at the river mouths.

The peptic digestion of the gastric extract was tested in five of the
fish from stomachs of which the *Bacillus coli communis* was obtained.
The extract was inactive in all, and showed only a trace of acidity. In
No. XII., 1895, the growth from the stomach proved to be a pure
cultivation of this organism.

Apart from the actual demonstration of the absence of food from the
stomach cavity in all the salmon examined, the slight acidity and small
digestive power of the extracts of the gastric mucous membrane recorded
lead to the conclusion that the fish both in the estuaries and in the
rivers were in a fasting condition.

(d) *The Nature of the Acid present in the Gastric Extract.*

Fifty ccm. of rectified spirit in which the chopped-up stomach of
Fish No. XXXIX., July, 1896, had been immersed, were distilled.

The first 40 ccm. which distilled over showed an acidity equal to
0·073 per cent. as HCl, but no hydrochloric acid was present.
Uffelmann's carbolo-ferric reagent was turned to a greenish-yellow hue.
The next 6 ccm. were neutral.

The remainder, 4 ccs., was shaken up with ether, and then an excess
of distilled water added. The water was removed from the ether by
means of a separation funnel. It amounted to 163 ccm. 100 ccs.
gave an acidity of ·00438 per cent. with no free hydrochloric acid, both
the phloroglucin-vanillin and the dimethyl-amido-azo-benzol tests
proving negative. On the addition of decinormal hydrochloric acid to
24 ccm. in the presence of dimethyl-amido-azo-benzol no reaction occurred
until 0·6 ccm. had been run in, or until 0·082 per cent. of the acid had
combined with the substances present, which were previously free from
such an acid combination. The addition of nitrate of silver in the
presence of nitric acid to the 100 ccm. tested for acidity caused a
precipitate weighing after incineration 0·025 gramme, 0·013 grammes
of which was soluble in nitric acid, leaving 0·012 grammes of chloride
of silver. This is equal to 0·00305 grammes of hydrochloric acid, or the
chlorine present was 0·003 per cent., while 0·013 grammes of silver,
equal to 0·0204 grammes of silver nitrate, were combined to organic
bodies. The ether contained 0·00219 per cent. of acid, and on
evaporation of part of the solution and addition of water Uffelmann's
reagent was discoloured, leaving a very slight yellow tinge.

Result, 50 ccs.

46 ccs. distilled; acidity equal to 0·0292 grammes HCl.
  4 ccs. remaining; „   „   0·00819   „   „
Watery extract 163 ccm., acidity equal to 0·0071 grammes HCl.
Ether extract   50 ccm.,   „   „   0·00109   „   „
Acidity of distillate,   0·0634 per cent.
Acidity of remainder,   0·2047 per cent.

Acidity of total 50 ccm., 0·07478 per cent. or 0·03739 gramme of acid.

The chlorine present in the form of chlorides or of hydrochloric acid only amounted to 0·00489 grammes or ·00978 per cent. as HCl, while 0·0148 grammes of hydrochloric acid had to be added before positive evidence of free mineral acid was obtained, or 0·0296 per cent. If the proportion of HCl, combining to proteid bodies be taken at 10 per cent. (Cf. *Journ. Anat. and Phys. Vol. XXVII., p. 195*) 0·148 grammes of proteid was present unattached to this acid.

Similarly it was shown that 0·0204 gramme of silver nitrate combined with organic material, which represents about 0·21 gramme of proteid.

From this stomach and its contents, therefore, no free mineral acid was obtained, but only a small quantity of combined hydrochloric acid and some organic acids.

The alcoholic extract of the chopped-up stomach of Fish No. XXVI., June 1896, caught in the upper waters, was mixed with ether, agitated, and then an excess of distilled water added. The water was separated from the supernatant ether and distilled to dryness; 0·00684 gramme of free hydrochloric acid was obtained in the distillate, and no organic acids were present.

The alcoholic extracts of the stomachs of three salmon caught in July 1895, a kelt, a fish which had been some time in fresh water, and a fish caught at the river mouth, showed no evidence of the presence of free hydrochloric acid, and were only slightly acid to litmus.

The gastric mucous membrane of salmon on their way to the spawning beds yielded no free hydrochloric acid to the alcohol (90 per cent.) in which it was immersed, but investigation showed the presence of some hydrochloric acid in combination with organic material. The alcoholic extract, however, contained a volatile organic acid which readily distilled over, and an organic acid which did not pass off on distillation. In one fish caught in the upper waters in June a small quantity of free hydrochloric acid was obtained from the alcoholic extract of the stomach, after distillation to dryness.

The almost constant coincidence of an increased acidity of the glycerine extract of the stomach with an enhanced digestive power must be regarded as being independent of the presence or absence of hydrochloric acid. Other acids are known to exert an appreciable power when acting in the presence of pepsine, and presumably, therefore, are capable of converting to some extent the forerunner of that ferment, pepsinogen, into active pepsine. The extracts were made from the chopped-up walls of the stomach and its mucoid contents, the acid present in the mucus and formed by bacterial action may have acted on the pepsinogen and converted part of it into pepsine. But the results of the bacteriological cultivations do not bear out this theory, as, with two exceptions, the fewer the organisms the greater was the power of the gastric extract.

Another difficulty, and a serious one, is afforded by the fact that the glycerine extracts, although made after immersion in alcohol for 48 hours or more, varied greatly in acidity, though without the presence of any bacterial decomposition. The peptic power with two exceptions corresponded with the acidity as determined.

It is impossible to even hazard any theory to account for the apparent paradox, and it only remains to state that the greater the acidity the fewer organisms, and the greater the peptic power of the ferment in the stomach, although the acidity of the extract may not be due to free hydrochloric acid.

5. TRYPTIC DIGESTION.

In a few instances the activity of the tryptic ferment of the

intestinal mucous membrane or of the pyloric appendages was estimated
by its power of digesting egg-albumin in 2 per cent. sodium carbonate
at the ordinary temperature of the room. The extracts were made in
the same way as those from the stomach. To prevent any fallacy from
the growth of organisms during the digestive action a small quantity
of chloroform was added to each—all the extracts possessed active
diastatic powers in alkaline solution.

One or two of the extracts were also used in an acid solution containing ·12 per cent. of hydrochloric acid. No action was observed.

I.—Intestinal Mucous Membrane.

The results have been tabulated—Table VII.—and show a much
greater proportion of digested albumin than in the peptic experiments.
In two fish the intestinal walls and contents were used, one of them a
kelt. Both were caught in July, and show a digestive power equal to
16·9 and 12·6 per cent. respectively :—

TABLE VII.

Estimation of the Proteolytic Action of the Glycerine Extract
of the Intestinal Mucous Membrane, and of the Pyloric
Appendages.

Fish Caught at the Mouth.

1. Intestine.

	Date.	Albumin Digested %.	Means.	Corresponding Proportion % Digested by Gastric Extract.
XI.	July, 1895	16·9	16·9 %	0·0

2. Appendages.

XXXIX.	December, 1895	42·9		—
XL.	„	28·1	(3) 43·6 %	7·9
XLI.	„	59·8		—
XIII.	May, 1896	51·0		10·4
XIV.	„	31·6	(10) 48·13%	0·0
XV.	„	54·4		33·3
XVI.	„	69·7	(7) 48·64%	22·2
XVII.	„	47·7		21·1
XVIII.	„	40·7		5·5
XIX.	„	45·4		1·1
10		48·13		12·4 Total Tested, 8

Fish Caught in the Upper Reaches.

1. Intestine.

	Date.	Albumin Digested %.	Means.	Corresponding
XIV. (Kelt)	July, 1895	12·6	12·6 %	0·0
Total, 12		42·56		9·95 Total Tested, 10

II.—Pyloric Appendages.

The extract from the appendages was much more active than that of
the intestine. Three fish caught in December afforded an average of
43·6 per cent. of albumin digested, and seven captured in May, one of
48·64 per cent. The greatest powers were shown by No. XVI. with
69·7 per cent., No. XLI. 59·8 per cent., No. XV. 54·4 per cent., No.
XIII. with 51·0 per cent., and No. XVII. with 47·7 per cent.

The corresponding figures for the peptic power in these fish show
that the July fish with no peptic power have the smallest proportion of

albumin digested by trypsin. Of the five which digested most albumin by the action of the appendical extract, four only were used for the estimation of peptic digestion. These four give an average amount of albumin digested by the extract from the appendages of 55·7 per cent., and by the gastric extract of 21·75 per cent. This last figure is much above the mean of 11·1 per cent. obtained from the total number of estimations of the peptic power.

The smallest amount of albumin digested (12·6 per cent.) was obtained from the action of the intestinal extract of a kelt caught in July, the next from that of an ascending fish in the same month (16·9 per cent.). The average percentage of albumin digested at the ordinary temperature by the extracts of the pyloric appendages was 48·13 per cent. This fact shows that a very considerable proteolytic power was still possessed by the secretion of the gland opening into the appendages. Dr. Gulland has shown that this is of the nature of a pancreas.

Although the intestines were invariably found to be empty, save for some mucus, the glandular elements in their walls, especially those in the mucous membrane of the appendages, contained a much more active zymogen than the gastric glands, and were able to afford a ferment, on the addition of an alkali, which was capable of a considerable amount of work.

5.—THE BACTERIOLOGY OF THE ALIMENTARY CANAL OF THE SALMON.

By A. LOCKHART GILLESPIE, M.D., F.R.C.P.Ed., F.R.S.E.

At one time it was seriously argued that the rapidity of digestion in fish is due to fermentative processes induced by bacterial forms. This theory was promulgated at a time when the large proportion of pepsine present in the gastric secretion of fish had not been ascertained. The fallacy presented by this theory is at once apparent when the time necessary for the bacterial decomposition of albuminous bodies, the bodies which are digested in the stomach, is considered; and especially when it is remembered that the temperature at which most bacteria grow is very much above that of the inner cavities in fishes.

The antiseptic action of an active and acid gastric juice has also been demonstrated, and the general rule may be laid down that the more active the gastric secretion and the greater its acidity the fewer bacteria can pass unharmed into the intestine.

Miescher Ruesch, in the paper referred to in a preceding section (II.), states categorically that the salmon caught in the upper reaches of the Rhine show less tendency towards early decomposition in the bowel, while those caught in tidal waters soon exhibit signs of putrefaction there. He adduces no detailed evidence in support of this statement, but suggests that the non-occurrence of decomposition in the intestines of salmon which have ascended a river is due to the fact that they do not feed, and therefore do not swallow bacteria with their food, arguing further that this absence of decomposition implies a self-imposed abstinence from food for some time before entering fresh water.

The observations detailed below were undertaken for the purpose of ascertaining the correctness of this statement, and to investigate the bacteriology of the alimentary canal in more detail. They include (1) an estimation of the number of organisms cultivated from the alimentary tract of the salmon; (2) a rough classification of these organisms; and (3) a comparison between their numbers and form in different parts of the rivers and at different times of the year.

1. MATERIAL USED.

Reference to Table I. shows that of the forty-one salmon employed twenty-nine were captured at the mouth, and twelve, one of these a kelt, in the upper reaches of the rivers:—

[TABLE.

TABLE I.

TABLE OF FISH EMPLOYED.

For Bacteriological Cultivations.

No.	Year.	Month.	No. of Fish.	Part of River.		Kelts.
				Mouth.	Upper.	
1	1895.	July.	X.		1	-
2	,,	,,	XI.	1	-	-
3	,,	,,	XII.	1	-	-
4	,,	,,	XIII.	1	-	-
5	,,	,,	XIV.		1	1
6	,,	August.	XXI.	1	-	-
7	,,	,,	XXII.	1	-	-
8	,,	,,	XXIII.	1	-	-
9	,,	September.	XXIV.	1	-	-
10	,,	October.	XXVII.	-	1	-
11	,,	,,	XXIX.	1	-	-
12	,,	,,	XXX.	1	-	-
13	,,	,,	XXXI.	1	-	-
14	,,	November.	XXXIII.	1	-	-
15	,,	,,	XXXV.		1	-
16	,,	,,	XXXVI.	1	-	-
17	1896.	March.	I.	1	-	-
18	,,	,,	II.	1	-	-
19	,,	,,	III.	1	-	-
20	,,	,,	IV.	1	-	-
21	,,	,,	V.	1	-	-
22	,,	,,	VI.	1	-	-
23	,,	,,	VII.	1	-	-
24	,,	May.	XXI.	-	1	-
25	,,	,,	XXIV.	1	-	-
26	,,	,,	XXV.	1	-	-
27	,,	June.	XXVI.		1	-
28	,,	,,	XXVII.	1	-	-
29	,,	,,	XXXVIII.	-	1	-
30	,,	,,	XXXIX.	1	-	-
31	,,	July.	XL.	1	-	-
32	,,	,,	XLII.		1	-
33	,,	,,	XLIII.		1	-
34	,,	,,	XLIV.	1	-	-
35	,,	,,	XLV.	1	-	-
36	,,	October.	LXIX.		1	-
37	,,	,,	LXX.		1	-
38	,,	,,	LXXI.	1	-	-
39	,,	November.	LXXVI.	1	-	-
40	,,	,,	LXXVII.	1	-	-
41	,,	,,	LXXVIII.	0	1	-
			Total, ...	29	12	1

Table II. (at p. 26 of the preceding section) further divides the total number into those caught at the river mouth and in the upper waters during different seasons of the year. All the fish captured in March came from the river mouths. During May and June four came from the lower, and three from the upper reaches. The July and August fish comprised nine from the river mouths, and four from the upper waters; of those captured from September to December nine came from the mouths and five from the higher parts.

2. METHODS.

In the case of each fish used for the purpose of determining the number and character of the micro-organisms present in the alimentary canal, " *stich* " cultivations were made on gelatine roll-tubes directly from the contents of the œsophagus, the stomach, and the intestine. As nearly as possible the same amount of material was, in every case, taken by means of a straight needle.   Second roll-tubes were inoculated from these in almost every case.

The colonies which appeared in the tubes were counted day by day after inoculation until their number became stationary, or so large as to be countless, or until liquefaction of the gelatine prevented any further enumeration.   When rapidly-growing and liquefying colonies rendered the gelatine fluid and inhibited the growth of the slower non-liquefying forms, the number of the latter was estimated from their colonies which began to grow before the liquefying growths had obscured their further development.

When the colonies were too numerous to count they were noted as being "in large numbers;" when still more numerous, and covering the whole surface of the gelatine with minute growths, they were termed "innumerable."   When it was wished to compare the relative numbers cultivated from different classes of fish, actual figures had to be substituted for these terms.   In each case 250 has been used to represent "a large number," and 500 to represent "innumerable."   The actual figures were certainly higher than those chosen, but those used indicate a somewhat definite proportion between the actual numbers present, and their constant substitution introduces no important element of error.

In this part of the research valuable help was rendered by Mr. Hume Patterson, the Laboratory Assistant.

I. Number of Micro-Organisms.

(a) *The Salmon Caught at the Estuaries.*

The chief details of the results obtained from the bacteriological investigation into the salmon caught in tidal waters are shown in Tables II. and III.   The results are grouped according to the season of the year at which the observations were made.   The number of the fish corresponding to that in the general Laboratory register is first given, and the year and month of capture are then noted:—

[Tables.

TABLE II.—TABULAR VIEW OF GENERAL RESULTS.

	March		May and June		July and August		September to November		Totals		Total	
	Month.	Upper.	Month.	Upper.	Month.	Upper.	Month.	Upper.	Month.	Upper.		
No Organisms in Canal,												
„ in Œsophagus,												
„ in Stomach,												
„ in Intestine,												
Only Moulds, in Œsophagus,												
„ „ in Stomach,												
„ „ in Intestine,												
Only Non-liquefying Organisms—												
„ in Œsophagus,												
„ in Stomach,												
„ in Intestine,												
Liquefying Organisms Grown—												
From all Parts,												
From Œsophagus,												
From Stomach,												
From Intestine,												
Bacillus Coli Communis—												
From all Parts,												
From Œsophagus,												
From Stomach,												
From Intestine,												

TABLE III.—MICRO-ORGANISMS GROWN FROM FISH.

(a) Fish caught at the Mouth of the Rivers.

No.	Year	Season	Bacteriological Examination. Tube I. Œsophagus	Tube I. Stomach	Tube I. Intestine	Tube II. Œsophagus	Tube II. Stomach	Tube II. Intestine
I.	1896	March	69 Yellow Liquefying	10 Green Moulds	—	—	—	8 Green Moulds, 1 Yellow Liquefying
II.	,,	,,	8 Green Moulds	12 Yellow Liquefying	2 Green Moulds, large number Yellow Liquefying (300)	1 Yellow Liquefying	—	4 Small Non-liquefying
III.	,,	,,	—	1 Green Mould	6 Non-liquefying	1 White Mould	—	—
IV.	,,	,,	8 Moulds	5 Moulds, 3 Yellow Liquefying	Liquefied; 100	—	1 Purple Mould	—
V.	,,	,,	2 Green Moulds	1 Purple Mould	—	—	—	—
VI.	,,	,,	1 Mould, 4 Yellow Liquefying	Large number (100) Yellow Liquefying	12 Yellow Liquefying	6 Moulds, 1 Yellow Liquefying	5 Moulds, 4 Yellow Liquefying	1 B. Coli Com.
VII.	,,	,,	3 Green Moulds	22 Moulds, 2 Yellow Liquefying	3 Green Moulds	—	7 Green Moulds	—
Total 2.	,,	,,	76 (22 Moulds, 54 Liquefying)	157 (29 Moulds, 117 Liquefying)	428 (5 Moulds, 6 Non-liquefying, 312 Liquefying)	8 (7 Moulds, 1 Liquefying)	18 (11 Moulds, 2 Liquefying)	14 (8 Moulds, 1 B. Coli Com.)
Average.	,,	,,	19·8 Colonies	19·5 Colonies	31·8 Colonies	1·1 Colony	1·8 Colony	2 Colonies

TABLE III.—*Continued.*

No.	Year.	Season.	Bacteriological Examination.					
			Tube I.			Tube II.		
			Œsophagus.	Stomach.	Intestine.	Œsophagus.	Stomach.	Intestine.
XXIV.	1896	May and June.	...	1 Liquefying	25 White Moulds	...	10 White Moulds	15 White Moulds
XXV.	,,	,,	1 White Mould	1 White Mould, 12 Non-liquefying	...	...	...	...
XXVII.	,,	,,	2 Green Moulds, Non-liquefying 4 Yellow Growths	3 Green Moulds	2 White Moulds, 2 Non-liquefying	2 Green Moulds, 1 Yellow Colony	2 Green Moulds	1 White Mould
XXXIX.	,,	,,	1 Green Mould	30 Green Moulds	...	1 Green Mould	4 Green Moulds	...
Total 4.	,,	,,	6 { 4 Moulds, 4 Non-liquefying	47 { 34 Moulds, 12 Non-liquefying, 1 Liquefying	29 { 27 Moulds, 2 Non-liquefying	4 : 3 Moulds	16 Moulds	16 Moulds
Average.	,,	,,	2 Colonies	11·7 Colonies	7·2 Colonies	1 Colony	4 Colonies	4 Colonies

TABLE III.—*Continued.*

No.	Year.	Season.	Bacteriological Examination.					
			Tube I.			Tube II.		
			Œsophagus.	Stomach.	Intestine.	Œsophagus.	Stomach.	Intestine.
XI.	1886.	July and August.	1 Coli, 250 Liquefying	250 Liquefying	250 Liquefying	2 Liquefying, 12 Non-liquefying	50 Yellow Liquefying, 3 Coli	1 Coli, 30 Liquefying
XII.	"	"	500 Non-liquefied	250 Coli	100 Coli, 250 Liquefying	1 Coli, 5 Non-liquefying	—	2 Liquefying
XIII.	"	"	500 Liquefied	500 Liquefied	—	3 Coli, 3 Non-liquefying, 4 Liquefying	3 Coli, 42 Liquefying	—
XXI.	"	"	500 Non-Liquefying	250 Liquefied	—	250 Non-liquefying	100 Liquefying	—
XXII.	"	"	1 Coli	1 White Mould	—	—	—	—
XXIII.	"	"	1 Coli, 8 Liquefying	3 Liquefying, 12 Non-liquefying	5 Coli, 16 Liquefying	—	—	—
XL.	1888.	"	250 Non-liquefying and Moulds	250 Non-liquefying	250 Non-liquefying	—	100 Non-liquefying	—
XLIV.	"	"	12 Liquefying	1 White Mould	250 Liquefying	—	—	1 Liquefying
XLV.	"	"	500 Liquefying	—	250 Non-liquefying	3 Moulds, 2 Liquefying, 4 Non-liquefying	—	—
Total 9.	"	"	2,523 { 1,279 Liquefying, 1,253 Non-liquefying	1,317 { 1,003 Non-liquefied, 314 Liquefying	1,371 { 766 Liquefying, 605 Non-liquefying	290 { 8 Liquefying, 282 Non-liquefying	295 { 102 Liquefying, 193 Non-liquefying	34 { 33 Liquefying, 1 Non-liquefying
Average.	"	"	280·3 Colonies	168·5 Colonies	152·3 Colonies	32·2 Colonies	32 Colonies	3·7 Colonies

TABLE III.—Continued.

No.	Year.	Season.	Bacteriological Examination.					
			Tube I.			Tube II.		
			Œsophagus.	Stomach.	Intestine.	Œsophagus.	Stomach.	Intestine.
XXIV.	1895.	September to November	—	1 Coli, 60 Liquefying	—	—	—	—
XXIX.	„	„	—	—	—	—	—	—
XXX.	„	„	—	—	—	—	—	—
XXXI.	„	„	—	—	16 Liquefying	—	—	1 Liquefying
XXXIII.	„	„	18 Non-liquefying Diplococcus	—	—	—	—	3 Coli, 1 Non-liquefying, 1 Yellow Torula
XXXVI.	„	„	1 Non-liquefying Diplococcus	1 White Mould	16 Non-liquefying	—	—	—
LXXI.	1896.	„	9 Yellow Liquefying	4 Non-liquefying	2 Liquefying, 50 Non-liquefying	—	—	—
LXXVI.	„	„	300 Non-liquefying	200 Non-liquefying	240 Liquefying	—	—	—
LXXVII.	„	„	250 Liquefying	—	250 Liquefying	—	—	—
Total 9.	„	„	579 { 259 Liquefying 319 Non-liquefying	560 { 60 Liquefying 500 Non-liquefying	584 { 518 Liquefying 66 Non-liquefying	0	0	0
Average.	„	„	64·2 Colonies	62·8 Colonies	63·7 Colonies	0	0	1 Colony

TABLE III.—*Continued.*

(b) Fish caught in the Upper Waters.

No.	Year.	Season.	Bacteriological Examination.					
			Tube I.			Tube II.		
			Oesophagus.	Stomach.	Intestine.	Oesophagus.	Stomach.	Intestine.
XXI.	1896	May & June	1 White Mould	—	2 White Moulds	1 White Mould	—	—
XXVI.	,,	,,	1 ,, ,,	2 White Moulds	1 Liquefying, 3 Non-liquefying, 1 White Mould	—	—	—
XXXVIII.	,,	,,	1 Liquefying	—	3 Non-liquefying	3 Non-liquefying	—	—
Total 3.	,,	,,	3 {1 Liquefying, 2 Non-liquefying}	2 (Moulds)	7 {3 Moulds, 3 Non-liquefying, 1 Liquefying}	4	0	0
Average	,,	,,	1	·66	2·3	1·3	0	0
X.	1896	July and August	250 Liquefying, 250 Non-liquefying	250 Liquefying, 250 Non-liquefying	100 Non-liquefying	25 Coli, 50 Liquefying	100 Coli, 100 Liquefying	—
XIV.	,,	,,	500 Non-liquefying	500 Non-liquefying, 200 Liquefying	200 Liquefying, 100 Non-liquefying	250 Non-liquefying	80 Non-liquefying, 18 Liquefying	8 Coli, 22 Liquefying
XLII.	1896	,,	300 Liquefying	300 Liquefying	250 Liquefying	—	—	—
XLIII.	,,	,,	350 Liquefying	350 Liquefying	250 Non-liquefying, 8 Liquefying	—	—	50 Non-liquefying
Total 4.	,,	,,	1400 {650 Liquefying, 750 Non-liquefying}	1400 {850 Liquefying, 550 Non-liquefying}	858 {728 Liquefying, 130 Non-liquefying}	325	396	80
Average	,,	,,	350 Colonies	350 Colonies	214·5 Colonies	82 Colonies	74·5 Colonies	20 Colonies

TABLE III.—*Continued.*

No.	Year.	Season	Bacteriological Examination.					
			Tube I.			Tube II.		
			Œsophagus.	Stomach.	Intestine.	Œsophagus.	Stomach.	Intestine.
XXVII.	1895	September to November	—	—	1 Yellow Sarcina	—	—	?
XXXV.	„	„	—	—	2 Coli, 12 Liquefying, 36 Non-liquefying	?	2 Non-liquefying	?
LXIX	1896	„	13 Liquefying, 162 Non-liquefying	356 Liquefying, 250 Non-liquefying	6 Non-liquefying, 3 Liquefying	?	?	?
LXX	„	„		39 Liquefying, 25 Coli, 70 Non-liquefying				
LXXVIII.	„	„	250 Liquefying	250 Liquefying	260 Non-liquefying			
Total 5.	„	„	423 { 263 Liquefying, 160 Non-liquefying	875 { 639 Liquefying, 346 Non-liquefying	321 { 55 Liquefying, 329 Non-liquefying			
Average	„	„	84·6 Colonies	175 Colonies	107·8 Colonies			

Entries are made in Tables II. and III. under the headings of Tube I. and Tube II. of the number and character of the micro-organisms grown from the three portions of the alimentary tract investigated, the œsophagus, stomach, and intestine. At the end of each of the four seasons into which the observations naturally fall, (1) March, (2) May-June, (3) July and August, and (4) from September to the end of the year, the approximate total number of growths is given, followed by the average per fish.

Similarly the total number of liquefying and non-liquefying organisms and the number of moulds are noted.

The results are graphically displayed in Charts I., II., and III.

In Chart I. the average number of organisms grown in the first roll-tubes is given for each season and for each section of the alimentary canal. The unbroken line represents the figures relating to the œsophagus, the broken line to those for the stomach, and the dot and dash line to the figures obtained from the intestine.

The method of representing the colonies when present in too large a number to admit of accurate computation has already been given. Moulds are included among the non-liquefying series owing to the comparatively long period which elapses before they affect the solidity of gelatine.

Chart II. has been constructed from the number of bacteria which were grown at each season from the different sections of the digestive canal, and which caused liquefaction of the gelatine medium.

The general form of the curves corresponds to that in the preceding Chart. The number of organisms grown from the œsophagus exceeds the number cultivated from the other two sections during the months of July and August in a marked manner. The chief point of difference between the figures for the organisms which liquefy gelatine and those for the total number lies in the larger proportion which of the former recorded as grown from the intestine.

It necessarily follows that the converse is true of the non-liquefying organisms, and a glance at Chart III. shows that fewer of these organisms were grown from the intestine and a greater number from the contents of the stomach. The figures for the œsophagus are practically the same in both cases.

The Charts show that the number of organisms present in the œsophagus exceeded those in the other sections of the alimentary canal in the latter half of the year. In March the numbers grown from the intestinal contents exceeded those from the stomach by an average of 12, and those from the œsophagus by 21. The numbers during this month were very low, while almost no growths were obtained from any of the second tubes. In fact, the total number of colonies which appeared in the second tubes inoculated from the three sections of the canals of seven fish only reached 35, or five colonies a fish, and 1·6 colonies on an average in each individual section. Still fewer growths were obtained from the salmon caught in May and June, an average of only two in each œsophagus, 11·7 in each stomach, and 7·2 in each intestine. The second tubes show a slightly larger number than in March, due to the number of moulds present in them.

In the case of the fish captured in July and August, large numbers of organisms were grown from each part of the canal. A glance at Chart I. shows that those cultivated from the contents of the œsophagus are largely in excess of the numbers present in the stomach or intestine. The numbers for the second tubes also show a great rise, except in those from the intestine. In the later months of the year the colonies grown fall below the numbers during July and August, but are still

PLATE II.

CHARTS I. to III.—The number of Colonies grown from the Alimentary Tract of Salmon caught in the Tidal Waters.

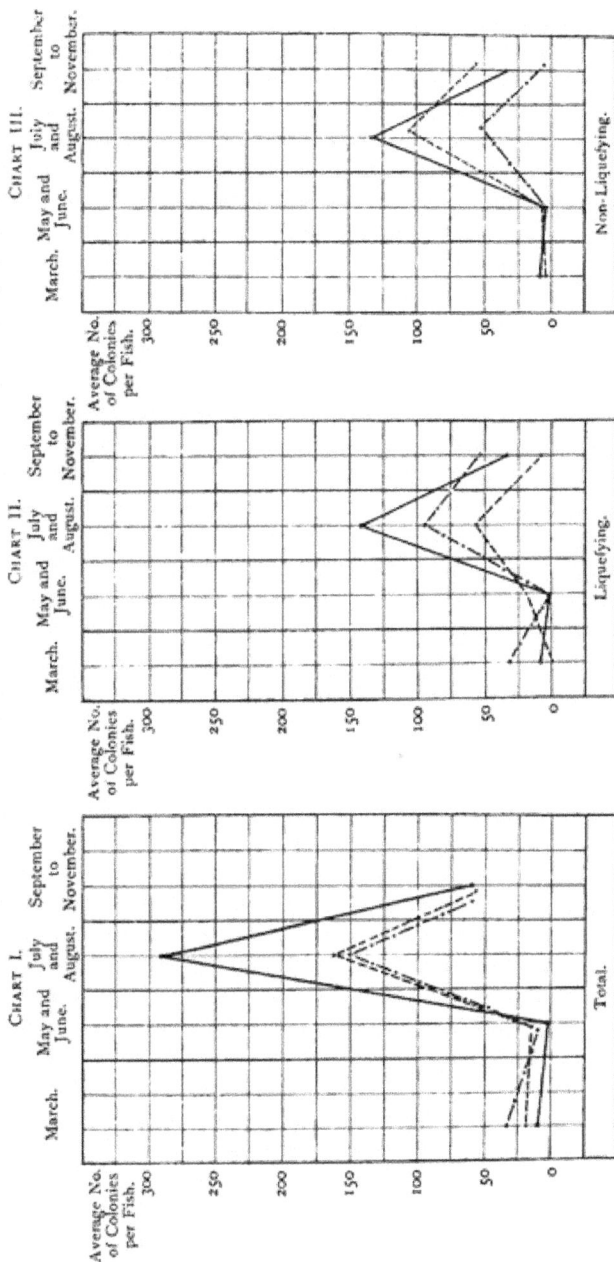

CHART I.

CHART II.

CHART III.

Total.

Liquefying.

Non-Liquefying.

March. May and June. July and August. September to November.

Average No. of Colonies per Fish.

300 250 200 150 100 50 0

———— In Œsophagus.   — — — — In Stomach.   ·—·—·—·— In Intestine.

PL

CHARTS IV. to VI.—The number of Colonies grown from the Alimentary Tract of Salmon caught in the Upper Waters.

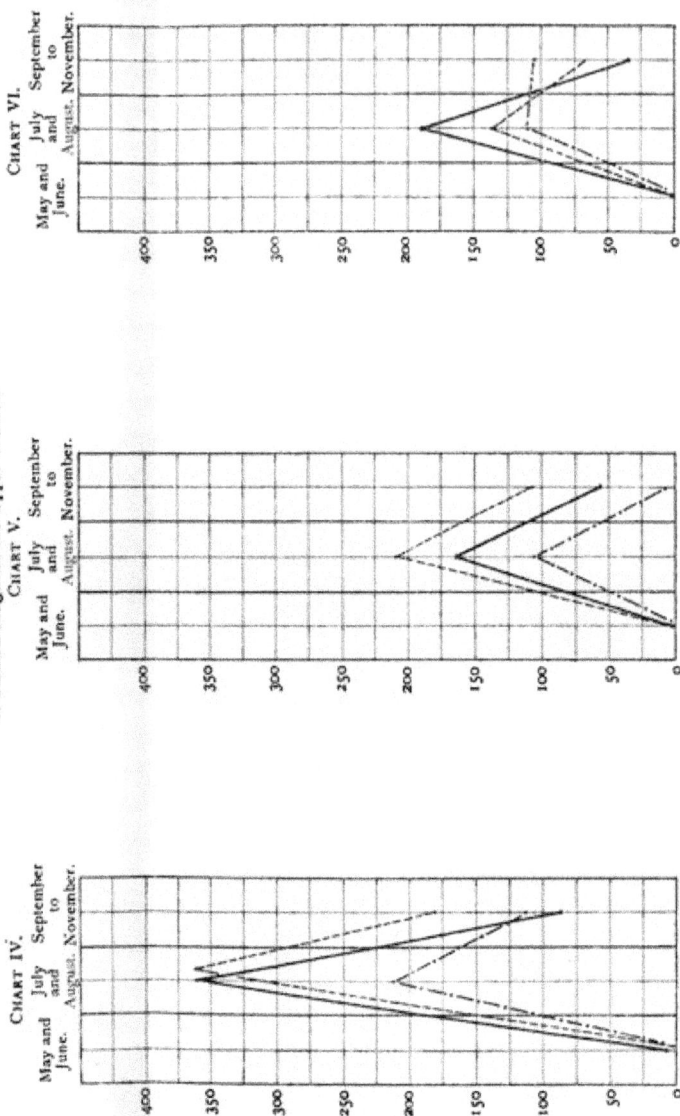

CHART IV.

CHART V.

CHART VI.

much more numerous than during the first half of the year. The mean numbers in all of the three sections of the alimentary canal are the same in those months when the results from the first roll-tubes are considered. And it is a most noteworthy fact that in not one of the six fish from which second tubes were made could a single growth be detected in the cultivations from both the œsophagus and stomach, while only six colonies in all, or one per fish, were obtained from the intestinal contents.

From these data it appears that there is a distinct connection between the temperature and the number of organisms present in the intestinal tract of the salmon caught at the mouth of fresh-water streams or in the tidal estuaries. The temperature of sea-water takes much longer to rise as the months advance than the temperature of the air, and the great increase in the number of bacteria cultivated during the months of July and August corresponds with the period of the year at which the temperature of shallow waters is near its maximum. The water in the rivers of Scotland, and in a considerable portion of the sea close to their mouths, must be affected in spring by the melted snow from the hills, with the result that the temperature of the tidal water remains low. Although the fish do not feed, they must swallow some of the water.

(b) *Salmon Caught in the Upper Waters.*

Twelve salmon from the upper reaches were employed for the investigation of the number of bacteria present in the alimentary canal. They fall to be divided as to the season of capture into three classes. Three were caught in May and June, four in July and August, and five in the months from September to November (Table III.)

The results are depicted in Charts IV., V., and VI. in a similar way to those obtained from the fish captured in tidal waters.

It is at once apparent that the same increase in the number of organisms occurred in the months of July and August, that almost no colonies were obtained in May and June, and that the bacteria found in the alimentary tract during the later months, though much less than in late summer, were more numerous than in the fish from the river-mouth (Chart IV.)

In May and June the total number of colonies which appeared in the roll-tubes was extremely small. Only twelve growths were cultivated from the alimentary tracts of three fish in the first tubes, or four apiece, and only four in the second tubes, all of which grew in the tubes inoculated from the œsophagus.

The increase in the numbers in July and August is shared equally by the œsophagus and stomach. The number grown from the intestinal contents was smaller by about three-fifths than the number in either of these sections.

In the autumn months the drop in the figures is most marked in the case of the œsophagus, and least in the case of the intestine.

The two **Charts** following (Charts V. and VI.), in which the organisms capable of liquefying gelatine have been separated from those incapable of so doing, show that the increase in the number of colonies obtained from the stomach contents in July and August is due to a predominance of bacteria able to dissolve gelatine. The number cultivated from the stomach exceeds that obtained from the œsophagus.

Practically no liquefying organisms were grown from any of the sections during the first season, or from the intestine in the third season of the year. During the last period, however, the colonies from the stomach contents were numerous, those from the œsophagus fewer, but still much above the number in the intestine.

The non-liquefying forms were obtained in numbers which, of course, coincide with the differences between the figures given in Charts IV. and V. Almost *nil* in May and June, the summer increase is most marked in the œsophagus, and proportionally in the intestine, when compared with the last chart; while the autumn decrease is greatest in the œsophagus, less in the stomach, and least in the intestine, from which, indeed, the same number of growths were obtained as in the months of July and August.

(c) Kelts. .

Cultivations were made from the intestinal tract of one kelt, caught in July 1895 in the upper reaches. Owing to the examination of only a single specimen of this class of fish, the results have been included in the preceding section.

A large number of non-liquefying growths were obtained from the œsophagus, of both liquefying and non-liquefying from the stomach and intestine, the latter section containing a few *bacilli coli communes*. The peptic activity of this fish was *nil*, the amount of albumen digested by an extract of the intestine, 12·6 per cent.

(d) Differences Between the Numbers of *Micro-Organisms Grown from the Alimentary Tract in Salmon caught in the Estuary and of those caught in the Upper Parts of the Rivers.*

In the 12 fish from the upper waters, the following numbers of organisms were counted :—

		From the Œsophagus.	Stomach.	Intestine.	Total.
Total,	..	1,826	2,277	1,419	5,522
Per Fish,	..	152	189	118	460

in the 29 caught in the estuaries—

Total,	..	3,185	2,267	2,197	7,649
Per Fish,	..	109·7	78·1	75·7	263·7

and in the total number of fish, *i.e.*, 41—

Total,	..	5,011	4,544	3,616	13,171
Per Fish,	..	122·2	110·8	88·4	321

In the total number of fish examined the œsophagus contained the largest number of organisms, the intestine the least.

On the other hand, the colonies grown from the stomach contents of the fish caught in the upper reaches were the most numerous, then the colonies from the œsophagus, and those from the intestine.

The colonies obtained from the œsophagus in fish from the estuaries presented a marked excess in numbers over those from the stomach and intestines. The numbers in the latter were nearly equal.

[TABLE.

Table showing the number of Micro-Organisms grown from different sections of the Alimentary Canal in Salmon caught in the Estuaries and in the Upper Waters:—

SEASON.	Number of Organisms per Fish.		
	Œsophagus.	Stomach.	Intestine.
July-August—			
From Upper Waters, . .	366	366	214·5
From Mouth, . . .	280·3	168·5	152·5
Difference, . . .	85·7	197·5	62·1
Difference per Cent. of Upper-Water Figures,	23·4	54·0	28·9
September-November—			
From Upper Waters, . .	84·6	177	110·8
From Mouth, . . .	64.2	62·8	63·7
Difference, . . .	20·4	114·2	47·1
Difference per Cent. of Upper-Water Figures,	24·0	64·0	42·5

The proportion between the number of colonies found in the tubes made from the œsophagus from July to November in the fish from the upper reaches and the mouths of the rivers differ only slightly. The number of colonies cultivated in autumn from the stomach contents of the fish from the upper waters exceeds that of the lower fish by 64 per cent. of its total, against 54 per cent. similarly found in the summer fish. In like manner in autumn the excess of organisms grown from the intestines of the upper-water fish over those found in the fish from the estuaries constitutes 42·5 per cent. of their number, compared with an excess of 28·9 per cent. from the same part of the canal in the summer fish.

The only season in which the figures for the average number of colonies grown from the alimentary canal of the salmon from the upper parts of the rivers are less than those for the fish from the mouths is that of May and June. During these months the excess in the number of organisms grown from the fish caught at the mouth over the number found in the others amounts to over 90 per cent. of the first number in all parts of the digestive tract.

We may further say, then, that in early summer the bacteria present in the alimentary tract of the salmon are less numerous when the fish is in the higher reaches of rivers than when it is still in tidal waters. On the other hand, the upper-water fish caught during the next two months are characterised by a much greater increase in the number of bacteria present in their alimentary canals, and consequently show a marked excess in actual numbers of bacteria over those in the fish at the river mouth. In the following months the rate of decrease in the number of bacteria present is greater in the fish in tidal waters, so that, although the organisms in the upper fish are not nearly so numerous as in the preceding periods, the proportional excess of bacteria in the upper-water fish over the number in those from the mouth is increased.

The number of organisms in the mucus in the œsophageal tube is practically the same in both sets of fish in May and June, but shows

D

in the upper-water fish an excess equal to about 24 per cent. of its total over the figure for the lower fish.

The colonies grown from the stomachs of the former class of fish were a little less numerous in May and June than in the latter, but show an enormous excess over those cultivated from the salmon caught at the mouth in July and August, and a proportionately larger excess in the next three months.

The figures for the colonies grown from the intestinal contents present the same features as those for the stomach, but in a less marked manner.

(e) *Relationship of Liquefying to Non-Liquefying Organisms.*

Turning to the figures given for the organisms which liquefy gelatine and for those which do not act on it in this way, when arranged in relation to season and site of capture, we see in Chart VII., (in which the red lines indicate the numbers of non-liquefying, the black lines those of the liquefying growths per fish), that the two curves representing the numbers of colonies grown from the œsophagus are almost identical. The curves for the colonies grown from the stomach show that in July and August the non-liquefying forms predominate, and that in the later months they still are the more numerous.

In July and August the liquefying growths are in the greater number in the intestine; while from September to November they become less numerous than the non-liquefying forms and moulds.

Chart VIII. shows more clearly the relative proportions between the two classes of organisms. In it the percentage values are depicted: that is, the percentage proportion of the liquefying organisms to the total number of growths obtained in each section of the alimentary canal, and at each period of the year. The black lines represent the result of the cultivations from fish captured in the upper waters; the red lines from those from the mouth.

Save in the lines for the intestine in the fish from the upper waters, and for the stomach in those from the mouths of the rivers, the general tendency is for the curves to rise as the year advances. In May and June liquefying organisms are practically absent, the 33·3 per cent. for the œsophagus of the upper-water salmon representing only one out of three growths. After a general rise in the percentage in July and August—a rise which is least in the stomachs of the fish from the mouth—the curves representing the proportions still tend slightly upwards. The œsophageal curve for the upper fish rises decidedly, the stomach curve for these fish remains the same, while that representing the colonies from the intestine in the fish from the mouth rises to 88 per cent. The lines which denote a fall are 1, for the œsophagus of the lower fish (and it is only slight), 2, for the stomach in the same fish, and 3, for the intestine of the other class as already mentioned.

When these curves are conjoined results are obtained (Chart IX.) which show more plainly the greater proportion of liquefying growths in the second part of the year. This proportion falls slightly in the œsophagus and stomach after July and August.

In March, during which no fish were received from the upper waters, the number of the liquefying organisms much exceeded that of the other class—383 to 53, or 87 per cent. of the total. They were most numerous in the intestine, least in the œsophagus.

Adding the totals together for both classes of organisms—totals which, it must be borne in mind, are largely empirical, we find that the organisms grown from 41 fish numbered 13,176, of which 6687 were moulds or did not liquefy gelatine, leaving 6489 organisms capable of dissolving

Percentage of Liquefying Organisms to Total from the Whole Tract.

May and June.   July and August.   September to November.

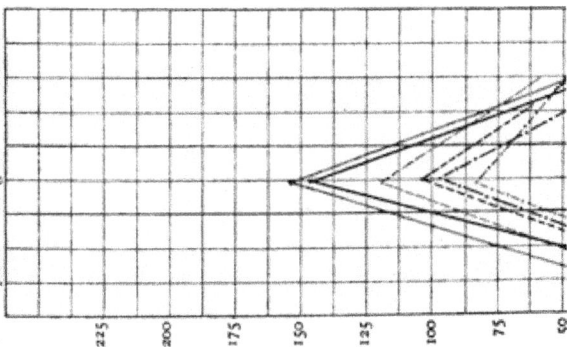

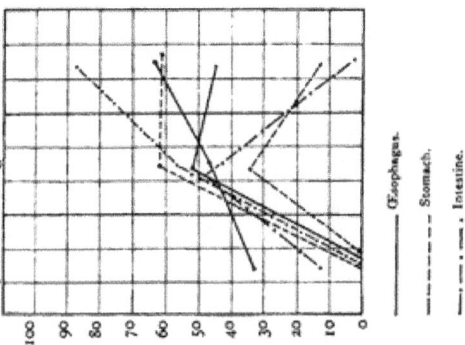

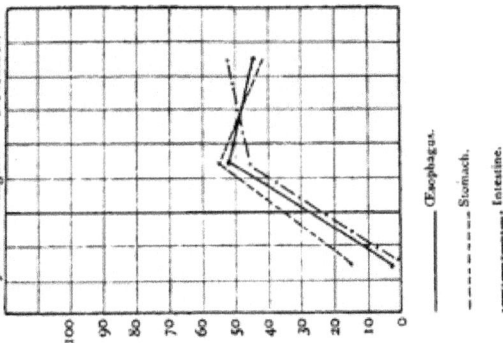

100 90 80 70 60 50 40 30 20 10 0

———— Œsophagus.
– – – – Stomach.
–·–·– Intestine.

Percentage of Liquefying Organisms to total in the Salmon from the River Mouths and the Upper Waters.

May and June.   July and August.   September to November.

100 90 80 70 60 50 40 30 20 10 0

———— Œsophagus.
– – – – Stomach.
–·–·– Intestine.

(Black) Fish from Upper Reaches.
(Red) Fish from Mouth.

Number of Organisms capable of Liquefying Gelatine and Non-Liquefying contrasted

May and June.   July and August.   September to November.

225 200 175 150 125 100 75 50

———— Œsophagus.
– – – – Stomach.
–·–·– Intestine.

(Black) Liquefying Colonies.
(Red) Non-Liquefying Colonies.

it. From each fish 321 colonies were obtained, or an average of 107 from each section of the alimentary canal.

In March the average number of colonies obtained from all parts of the tract was 62; in May and June, 13·7; in July and August, 697; and September to November, 255. It is difficult to account for the larger number obtained from the fish caught at the river mouths in March than from those captured during May and June, especially as during the latter month several of the fish came from the higher reaches.

CONCLUSIONS.

From these observations it would appear that the alimentary tract of the salmon in tidal waters, preparatory to ascent of the rivers, contains a smaller number of bacteria than the tract of those fish which have proceeded up the streams. The number of growths obtained from the œsophagus of the lower fish is below that found in the upper fish (27 per cent.), from which fact it may be concluded that fewer organisms are swallowed, but that the difference is not very great. The greatest contrast between the results for the two classes of fish is presented by the figures for the growths cultivated from the stomachs. Fewer colonies were obtained from the stomachs of the lower-water fish than of the upper, while the liquefying bacteria formed the larger part of the colonies in the case of the upper fish.

Miescher Ruesch affirms the exact opposite. He finds that the salmon, from the upper reaches do not decompose so quickly as the lower fish; due, he suggests, to the small number of organisms swallowed by the upper fish while fasting. Direct experiment shows that the putrefactive bacteria are more numerous in the upper fish than in the lower, as well as affording evidence of the greater number of all bacterial forms in these fish. Although the upper-water salmon may not feed, they must swallow some of the water of the streams more or less frequently. The variations observed are most probably due to the proportion of organisms in the surrounding water—more numerous in the fresh water than in the tidal waters, and consisting of a greater number of liquefying or putrefactive forms. That the diminution in the numbers of organisms in the alimentary canal below the œsophagus is much more marked in the lower fish than in the upper cannot be due to any increased effect of gastric digestion, is shown by the smaller percentage of albumin digested by their extracts. The acidity, however, of the gastric extracts of the lower fish is slightly in excess of the acidity found in the upper salmon; while the number of non-liquefying colonies, although actually less numerous, are, relatively to the total, in a much higher proportion.

These facts seem to lead to the following conclusions:—

1. Fewer organisms are swallowed by salmon in tidal waters.

2. A larger proportion of these organisms are non-putrefactive and acid-forming.

3. The presence of these non-putrefactive organisms in excess of the liquefying forms prevents the rapid growth of the latter.

4. As the members of the putrefactive class of bacteria grow much more quickly than the bacteria forming the other class, fewer colonies can be obtained from each part of the canal.

5. The proportion between the total number of organisms grown from the stomach and intestine in the lower and upper fish did not differ so much in the warm summer months as during the late autumn, when the colonies, grown from these sections in the upper fish, largely exceeded the number grown from those in the fish caught at the mouth.

6. Of all the fish examined, the organisms were very much more

numerous in those caught in July and August than in those caught in spring, or even in May and June. The autumn fish contained a greater number than the spring, but fewer than the summer fish.

2. GENERAL CHARACTERS OF THE MICRO-ORGANISMS CULTIVATED.

(a) *Absence of Micro-Organisms.*

In three fish, all caught in October (one in the upper waters) no organisms appeared in any of the tubes:—

TABLE IV.

GIVING THE RESULTS OF THE BACTERIOLOGICAL CULTIVATIONS.

1. *No Organisms Grown from the Alimentary Canal.*

No.	Date.	Part of River.
1	XXVII.	24th October, 1895.	Upper.
2	XXIX.	28th October, 1895.	Mouth.
3	XXX.	28th October, 1895.	Mouth.

2. *No Organisms Grown from—*

	Œsophagus.				Stomach.				Intestine.		
	No.	Date.	Part of River.		No.	Date.	Part of River.		No.	Date.	Part of River.
1	XXIV.	1895 Sept. 4	Mouth.	1	XXXI.	1895 Oct. 28	Mouth.	1	XIII.	1895 July 25	Mouth.
2	XXXI.	Oct. 28	,,	2	XXXIII.	Nov. 9	,,	2	XXI.	Aug.17	,,
3	XXXV.	Nov. 24	Upper.	3	XXI.	1896 May 26	Upper.	3	XXII.	Aug.23	,,
4	XXIV.	1896 May 29	Mouth.	4	XXXVIII.	June 9	,,	4	XXIV.	Sept. 4	,,
5	LXIX.	Oct. 27	Upper.	5	XLV.	July 18	Mouth.	5	XXXIII.	Nov. 9	,,
				6	LXXVII.	Nov. 7	,,	6	L.	1896 Mar. 6	,,
								7	V.	Mar. 6	,,
								8	XXV.	May 30	,,
								9	XXXIX.	June 10	,,

3. *Only Moulds Grown from—*

	Œsophagus.				Stomach.				Intestine.		
	No.	Date.	Part of River.		No.	Date.	Part of River.		No.	Date.	Part of River.
1	II.	1896 Mar. 6	Mouth.	1	XXII.	1895 Aug.23	Mouth.	1	VII.	1896 Mar. 6	Mouth.
2	III.	,,	,,	2	XXXVI.	Nov. 23	,,	2	XXIV.	May 29	,,
3	IV.	,,	,,	3	I.	1896 Mar. 6	,,	3	XXXVIII.	June 9	Upper.
4	V.	,,	,,	4	III.	,,	,,	4	XXI.	May 26	,,
5	VII.	,,	,,	5	V.	,,	,,				
6	XXI.	May 26	Upper.	6	XXVI.	June 2	Upper.				
7	XXV.	May 30	Mouth.	7	XXVII.	,, 3	Mouth.				
8	XXVI.	June 2	Upper.	8	XXXIX.	,, 10	,,				
9	XXXIX.	June 10	Mouth.	9	XLIV.	July 16	,,				

TABLE IV.—*Continued.*

4. *No Bacteria Liquefying Gelatine Grown from—*

(Non-liquefying forms, with or without moulds, present.)

	Œsophagus.				Stomach.				Intestine.		
	No.	Date.	Part of River.		No.	Date.	Part of River.		No.	Date.	Part of River.
1	XII.	1895 July 25	Mouth.	1	XII.	1895 July 25	Mouth.	1	X.	1895 July 25	Upper.
2	XIV.	,,	Upper.	2	XXXV.	Nov.24	Upper.	2	XXXVI.	Nov. 23	Mouth.
3	XXI.	Aug.17	Mouth.	3	XXV.	1896 May 30	Mouth.	3	XXXV.	,, 24	Upper.
4	XXII.	,, 23	,,	4	XL.	July 9	,,	4	III.	1896 Mar. 6	Mouth.
5	XXXIII.	Nov. 9	,,	5	LXXI.	Oct. 27	,,	5	XXVII.	June 3	,,
6	XXXVI.	,, 23	,,	6	LXXVI.	Nov. 7	,,	6	XLV.	July 18	,,
7	XXVII.	1896 June 3	,,					7	LXXVIII.	Nov. 7	Upper.
8	XL.	July 9	,,					8	XL.	July 9	Mouth.
9	LXXVI.	Nov. 7	,,								

5. *Organisms similar to, or identical with, Bacillus Coli Communis.*

	Œsophagus.				Stomach.				Intestine.		
	No.	Date.	Part of River.		No.	Date.	Part of River.		No.	Date.	Part of River.
1	X.	1895 July 25	Upper.	1	X.	1895 July 25	Upper.	1	XI.	1895 July 25	Mouth.
2	XI.	,,	Mouth.	2	XI.	,,	Mouth.	2	XII.	,,	,,
3	XII.	,,	,,	3	XII.	,,	,,	3	XIV.	,,	Upper.
4	XIII.	,,	,,	4	XIII.	,,	,,	4	XXIII.	Aug.31	Mouth.
5	XXII.	Aug.23	,,	5	XXIV.	Sept. 4	,,	5	XXXVI.	Nov.23	,,
6	XXIII.	,, 31	,,	6	LXX.	1896 Oct. 27	Upper.	6	VI.	1896 Mar. 6	,,
								7	LXIX.	Oct. 27	Upper.

6. *Micro-organisms which Liquefied Gelatine were Grown from—*

1. IN ALL PARTS.

	No.	Date.	Part of River.
1	XI.	25th July, 1895.	Mouth.
2	XXIII.	31st August, 1895.	,,
3	VI.	6th March, 1896.	,,
4	XLII.	14th July, 1896.	Upper.
5	XLIII.	16th July, 1896.	,,
6	LXX.	27th October, 1896.	,,

TABLE IV.—*Continued.*

2. ONLY IN—

	Œsophagus.			Stomach.			Intestine.		
	No.	Date.	Part of River.	No.	Date.	Part of River.	No.	Date.	Part of River.
1	X.	1895 July 25	Upper.	X.	1895 July 25	Upper.	XII.	1895 July 25	**Mouth.**
2	XIII.	,,	Mouth.	XIII.	,,	Mouth.	XIV.	,,	Upper.
3	I.	1896 Mar. 6	,,	XIV.	,,	Upper.	XXXI.	Oct. 28	Mouth.
4	XXXVIII.	June 9	Upper.	XXI.	Aug.17	Mouth.	II.	1896 Mar. 6	,,
5	XLIV.	July 16	Mouth.	XXIV.	Sept. 4	,,	XXVI.	June 2	Upper.
6	XLV.	,, 18	,,	II.	1896 Mar. 6	,,	XLIV.	July 16	Mouth.
7	LXXI.	Oct. 27	,,	IV.	,,	,,	LXIX.	Oct. 27	Upper.
8	LXXVII.	Nov. 7	,,	VII.	,,	,,	LXXI.	,,	Mouth.
9	LXXVIII.	,,	Upper.	XXIV.	May 29	,,	LXXVI.	Nov. 7	,,
				LXIX.	Oct. 27	Upper.	LXXVII.	,,	,,
				LXXVIII.	Nov. 7	,,	IV.	Mar. 6	,,

No growths were obtained from the œsophagus in five fish—four of which were caught during the later months of the year, and one in May. Two of the fish, during the autumn months, were from the upper reaches. Similarly the tubes were sterile which were made from the stomach in six fish—three from the upper waters in November, May, and June, three in October and November from the mouth. The nine fish from whose intestines no growth was obtained were all caught at the mouth —two in March, one in May, one in June, one in July, two in August, one in September, and one in November.

(b) *Moulds.*

In many instances the colonies observed were entirely made up of moulds: thus in nine cases from the œsophagus, in the same number from the stomach, and in four from the intestine, this class of organism was present alone.

In the fish caught at the river mouths in March the appearance of moulds, apart from other growths, was frequent: thus in the œsophagus of five fish, in the stomach of three, and in the intestine of one caught during March, only moulds developed in the tubes. The majority of the other instances of such single growth occurred in May and June, only three belonging to a later period of the year. All the five fish which belong to this class (from the upper reaches) were captured in June.

(c) *Absence of Liquefying Bacteria.*

No bacteria capable of liquefying gelatine were grown from different sections of the alimentary tract in several fish. This class includes the cultivations in which non-liquefying bacteria, either alone or along with moulds, were present.

In one fish caught at the river mouth, in July 1896, no liquefying colonies were observed in any of the tubes made from its alimentary tract. The œsophageal contents were free from liquefying colonies in nine fish, one of them from the upper water; the gastric contents in six fish, one of them from the upper water in November 1895; and the intestinal contents in eight, with three from the same part of the rivers —one in July and two in November. The fish included in this class may be divided into:--

	Œsophagus.	Stomach.	Intestine.	Total.
March,	—	—	1	1
May,	—	1	-	1
June,	1	—	1	2
July,	3	2	3	8
August,	2			2
October,	-	1	-	1
November,	3	2	3	8
Total,	9	6	8	23 occasions.

(d) *Bacillus Coli Communis.*

Among the growths which did not liquefy gelatine there were a number closely resembling in the form of their colonies and in their microscopic appearance the *Bacillus coli communis*. It is now recognised that several organisms may be included under this name. The individual differences between the varieties are very slight.

The *Bacillus coli communis* was found in the œsophagus and in the stomach in six fish, in the intestines in seven. In two of these (which were caught at the river mouth in July) the organism was detected in each section of the digestive canal. In many of these other organisms were also present.

All the fish from which this bacillus could be cultivated from the œsophagus were caught during July and August; and, in four of the fish caught during these months, from the stomach and intestine. The other occasions on which it was present were during the later months—except for one in March. Only four of the fish came from the upper waters.

3. SPECIAL CHARACTERISTICS OF SOME OF THE MICRO-ORGANISMS CULTIVATED FROM THE CONTENTS OF THE ALIMENTARY CANAL.

(a) *Organisms which Liquefy Gelatine.*

The varieties of bacteria capable of liquefying gelatine which were found in the mucus of the alimentary canal may be divided into one or two types:—

1. Bacilli liquefying gelatine in "stich" cultures, and producing a

dirty yellow-red deposit at the bottom of the tube, very similar in appearance to the sediment in urine due to amorphous urates. The cultures soon acquired a disagreeable odour. Under the microscope long thin motile bacilli were seen, many of them in pairs, or chains ; very similar to the " roter " bacillus found in water.

This class of organism was confined entirely to the salmon sent in for examination in July and August, 1895.

The actual fish from which they were obtained were as follows :—

	Œsophagus.	Stomach.	Intestine.
Fish from the Mouth - -	XIII. (1895), July. XXIII. (1895), August.	XXI. (1895), August.	XI. (1895), July. XII. (1895), August,
Fish from the Upper Water	XIV. (1895), July.	XIV. (1895), July.	XIV. (1895), July.

2. Micro-organisms liquefying gelatine and forming a white deposit at the lower end of the tubes.

In one or two instances these organisms were further examined. In No. XIII., July 25, 1895, very freely motile bacilli, their length in relation to their breadth being $2\frac{1}{2}$ as to 1, were found in the stomach. They frequently were present in pairs. The gelatine was very rapidly liquefied, and rendered strongly alkaline.

Organisms rapidly liquefying gelatine and forming a white flocculent deposit were grown from :—

Œsophagus.	Stomach.	Intestine.
X. (1895) July.	XIII. (1895) July.	XXXI. (1895) October.
XIII. „ „	XIV. „ „	XXVI. (1896) June.
V. (1896) March.	XXIV. „ Septr.	XLII. „ July.
XXXVIII. „ June.	XXIV. (1896) May.	XLIII. „ „
XLII. „ July.	XLII. „ July.	XLIV. „ „
XLIII. „ „	XLIII. „ „	LXXI. „ October.
XLIV. „ „	LXXVIII. „ Novr.	LXXVII. „ Novr.
XLV. „ „		
LXXI. „ October.		
LXXVII. „ Novr.		
LXXVIII. „ „		

3. Organisms colouring the liquefied gelatine light yellow, and forming a bright yellow deposit:—

Œsophagus.		Stomach.		Intestine.	
XIII.	(1895) July.	XI.	(1895) July.	VI.	(1896) March.
VI.	(1896) March.	VI.	(1896) March.	LXIX.	,, October.
LXX.	,, October.	VII.	., ,,	LXX.	,, ,,
		LXIX.	., October.		
		LXX.	,, ,,		

4. Large dense creamy colonies slowly liquefying the gelatine into holes with sharp edges, no pigment formed. Very minute bacteria :—

XI.	XXIII.	XI.
XXIII.		XXIII.

5. Bacilli liquefying gelatine and forming a light yellow deposit without colouring the supernatant fluid :—

I.	X.	II.
	II.	IV.
	IV.	

(b) Organisms not Liquefying Gelatine.

1. Bacillus coli communis (cf. Table IV., 5, p. 53).

2. Small round yellow colonies, formed by short bacteria, often in pairs, with rounded ends, very like micrococci, some in zoogloea. The outer ends of those joined in pairs stain a darker colour. The pigment formed was very bright in colour, and the growths were raised above the surface of the gelatine in a similar manner to those of yeasts.

This bacterium was grown from the œsophagus of Fish No. XII., July 1895, No. XXVII., June 1896, No. LXX., October 1896 ; from the stomach of Nos. XXIV. and XXV., May 1896, No. LXX., October 1896 ; and from the intestine in Fish No. VI., March 1896, No. XXVI., June 1896, No. LXIX. and No. LXX. in October 1896.

3. Small round yellow colonies of cocci, which were usually present in the form of diplococci ; the growths were raised on the surface, and like nail heads.

These diplococci were only found on two occasions, both during the month of November 1895, in the œsophagus of No. XXXIII. and of No. XXXVI. Both these fish were caught in the lower waters.

4. Organisms forming creamy moist circular colonies on gelatine. The growths were white, faintly tinged with yellow in colour, raised on the surface of the gelatine, circular and moist. The organisms present were small non-motile bacteria, many of them looking like dumb-bells. They were found along with colonies of a very similar appearance, but which liquefied the gelatine, in the œsophagus of Fish No. XI., 1895, and in

the stomach of Fish No. XXV., the intestine of No. XXVII., and the œsophagus of No. XXXVIII., 1896.

5. Small white colonies whose nature was not further determined. These occurred in a great number of the tubes.

(c) *Yeasts, Moulds, and Sarcinæ.*

Very few organisms akin to the yeasts were found. In No. 36, November 1895, a yellow torula was grown from the intestine, a yellow sarcinæ from the intestine of No. 35, in the same month, and in the stomach of No. 70, October 1896, 20 yellow yeasts or torulæ were found.

The members of the mould class were much more numerous. Table IV. gives a list of the sections in which moulds were the only organisms grown. In the case of the œsophagus, cultivations of moulds alone were almost entirely confined to the late autumn months, in the stomach to **three out** of six the same applies, **while in** the intestine the season has little to do with their appearance, **two** occurring in March, two in May **and June, three** in July and August, and two in the later months. Five of these fish were caught in the upper water.

Examination of **Table V.**, giving some of the results of the bacteriological **investigations in** the fish caught at the river **mouth**, shows that **moulds were most** numerous in the fish caught in March, May, and June :—

TABLE V. (See also Table III).

		Œsophagus.	Stomach.	Intestine.	Total.
March, . .	7 Fish.	22 Moulds. 16 Greenish. 4 White.	20 Moulds. 11 Greenish. 8 White. 1 Purple.	5 Moulds. 5 Greenish.	47
May and June,	4 Fish.	4 Moulds. 4 Greenish.	34 Moulds. 33 Greenish. 1 White.	27 Moulds. All white.	65
July and Aug.,	9 Fish.	30 Moulds. White.	2 Moulds. White.	None.	32
September to November,	9 Fish.	None.	1 White mould	None.	1
	29	56	57	32	145

Similarly the fish caught in the upper water gave results as follows :—

		Œsophagus.	Stomach.	Intestine.
May and June -	3 Fish.	2 White moulds.	2 White moulds.	3 White moulds.
July and August	4 Fish.	None.	None.	None.
September to November .	5 Fish.			
	12	2	2	3 = 7

Moulds were therefore much more common in the fish caught in tidal waters, and in the earlier part of the year, than in those from the upper reaches or captured in autumn. The number found in the œsophagus and stomach were much in excess of those grown from the intestinal contents.

(d) *Micro-Organisms in Two Series of Fish.*

An interesting series of cultivations was obtained from a group of fish sent in on the 5th of November 1896, from the mouth of the Spey. Their numbers are 76, 77, and 78 :

	Œsophagus.	Stomach.	Intestine.
No. 76 -	*a.* 300 non-liquefying spreading colonies.	*b.* Innumerable small colonies, a pure growth. Non-liquefying.	*c.* Gelatine rapidly liquefied. White deposit.
No. 77 -	*d.* Gelatine rapidly liquefied with a white deposit.	*e.* No growths.	*f.* As in œsophagus.
No 78 -	*h.* As in No. 77.	*i.* As in œsophagus.	*j.* Innumerable, no liquefying growths. Pure cultivation.

The organisms in *c, d, f, h,* and *i* were identical. They liquefied gelatine, forming a pure white precipitate, and consisted of small motile bacilli, some of them joined together in chains.

The growths in *b* and *j* were also identical. Innumerable small pin-head colonies, not liquefying the gelatine, and showing no variations when grown in second dilutions. The organisms were very short, small, plump bacteria, almost like cocci, immotile, and all of the same size.

No growth appeared in *e.*

The organism present in *a* was a short, broad bacterium, in many instances gathered into zoogloeæ, in others prolonged into long jointed filaments. It was non-motile and did not liquefy gelatine.

In these three fish caught at the same place and on the same date surprising variations in the number and nature of the organisms occur. No rule seems to guide either the number or the forms present, either when one fish is contrasted with another, or when succeeding sections of the alimentary canal are compared.

Another interesting group of fish was sent in on October 27, 1896. No. 69, from the upper Helmsdale ; No. 70, from the upper Dee ; and No. 71, from the mouth of the Dee :—

[TABLE.

	(Esophagus.	Stomach.	Intestine.
	a.	*b.*	*c.*
No. 69 - -	None.	Innumerable. Yellow liquefying and non-liquefying.	2 *coli communis* ; 12 bright yellow liquefying ; 30 opaque yellow non-liquefying.
	d.	*e.*	*f.*
No. 70 - -	13 bright yellow liquefying. 60 yellow non-liquefying and 100 minute colonies, probably the same as last.	25 *coli communis.* 30 yellow liquefying. 20 yellow yeasts. 50 small yellow non-liquefying.	9 small yellow liquefying.
	g.	*h.*	*i.*
N 71 - -	9 liquefying with a white deposit.	4 small white non-liquefying.	2 liquefying with a white deposit. 50 small white non-liquefying.

In the two fish from the upper reaches, tubes *b, c, d, e,* and *f* showed the presence of a liquefying organism, a small, short, and thin bacillus, which coloured the fluid portion of the medium a bright yellow, and produced a deposit of a similar colour. The third fish presented none of this variety. The organism which did not liquefy the gelatine in *b, c, d,* and *e* also gave a yellow colour to its growths. Its characters have already been described. In *c* and *e* a few colonies of the *Bacillus coli communis* were found, and in *e* some yellow yeasts.

No. 71 possessed no organism producing pigment, and the number of colonies cultivated from it were small in number.

CONCLUSIONS.

No very definite statement can be made in connection with the character of the organisms cultivated. The characters of the growths varied much more than the total numbers. The entire absence of growths from the whole tract was noted in three fish, all caught in October, and from sections of the tract on 20 occasions in 17 fish, only four of which were caught during the months of July and August.

Similarly no organisms, save moulds, were grown from 22 separate sections in 16 fish, only two belonging to fish caught during these months.

These two facts point to the frequent absence of organisms, other than moulds, from the water during the greater part of the year.

Forms of bacteria capable of liquefying gelatine by their growth were absent in eight sections out of a possible 30 in fish caught in July, in two out of nine in August, and eight out of 21 sections in November In March and October they were seldom absent.

Cultivations consisting entirely of non-liquefying forms were most common in November and July.

Of the total cultivations, containing non-liquefying forms alone, five only were from upper-water fish, and of these two were caught in July and August.

That is to say, that out of 36 cultivations made from the segments of the alimentary tract of fish from the upper water, five, or nearly 14 per cent., were free from liquefying growths, and of these, two occurred in July out of 12 tested, or 16 per cent. Of the 87 cultivations made from the fish from the mouth, 18 developed no liquefying colonies, or 20·7 per cent. Of the 18, eight were obtained in July and August, out of 27 cultivations, or about 29 per cent.; six in late autumn out of 27, or 22 per cent.; three in May and June out of 12, or 25 per cent.; and one in March out of 21.

But to obtain the correct figures, the number of occasions on which no growths or moulds only were present should be deducted from the total numbers. Doing this we find that no liquefying organisms were found in :—

	March.	May and June.	July and August.	September to November.
Upper Water	—	—	2 in 12 (16·6%)	3 in 11 (27·2%)
Lower Water	1 in 10 (10%)	3 in 4 (75%)	8 in 21 (38%)	6 in 18 (33·3%)

Thus it is clear that *colonies of bacteria which liquefy gelatine and cause putrefaction are much more frequently absent in the fish caught in the tidal waters than those caught in the upper reaches of the river.*

If the numbers of the cultivations in which no organisms, only moulds, or only non-liquefying forms, developed, be added together, we find the following :—

	March.	May and June.	July and August.	September to November.
Upper Water	—	7 in 9 (77·7%)	2 in 12 (16·6%)	7 in 15 (46·6%)
Lower Water	12 in 21 (57·1%)	11 in 12 (91·6%)	14 in 27 (51·6%)	15 in 27 (55·5%)

A total in the upper-water fish of 16 in 36, or 44·4 per cent., and in the lower of 52 in 87, or nearly 60 per cent.

Putrefactive organisms are therefore less common in the alimentary tract of salmon living in tidal waters throughout the year, while they are almost entirely absent in fish caught in May and June either at the mouth or the upper reaches of rivers.

The *Bacillus coli communis* was grown on 19 occasions, 14 of these occurring in July and August. Five cultivations from fish caught in the upper waters contained this organism.

On several occasions fish caught on succeeding days, or a short time after one another, afforded evidence of the presence of the same organism. Thus the same diplococcus was given from two fish caught in November 1895.

Moulds are more common in the alimentary tract of fish living in tidal waters, especially during the earlier part of the year.

The two series, each of three fish, which are separately noted, show that the bacteriological conditions may be very different in salmon caught about the same date. In the first series the three fish were captured on the same day and at the same place, but the colonies grown from them differ in both character and distribution. In the second case two of the fish came from the same river, one from the mouth and one from the upper part, the third from the upper part of another stream; yet the two fish from the upper reaches contained an identical small liquefying organism, and the *Bacillus coli communis*, neither of which was grown from the third fish from the river mouth.

The two main factors which influence the number and nature of the bacteria in the alimentary tract of the salmon when in tidal waters preparing to ascend fresh-water streams, or during this ascent, appear to be,—1, the season of the year, including of course in this the influence of the temperature of the water, and, —2, the number and nature of the organisms present in the water. To suggest that organisms swallowed with food can cause an increased tendency to early decomposition appears to be not in accordance with the fact that the gastric juice of fish when feeding contains much more acid than is required to destroy or inhibit all organisms, especially those putrefactive forms which are easily effected by the agent. When fasting the acidity is not sufficient to kill the organisms which enter with the water swallowed; when feeding, many of the organisms fail to survive the action of the acid in the secretion of the gastric glands. Tidal water must contain fewer organisms capable of living in the alimentary tract than fresh water.

B. CHANGES IN THE WEIGHT AND IN THE CONSTITUENTS OF THE MUSCLES, GENITALIA, AND OTHER ORGANS DURING THE SOJOURN OF THE SALMON IN FRESH WATER.

6.—CHANGES IN WEIGHT AND CONDITION OF SALMON AT DIFFERENT SEASONS IN THE ESTUARIES AND IN UPPER REACHES OF THE RIVERS.

By D. NOËL PATON, M.D., F.R.C.P.Ed.,

AND

JAMES C. DUNLOP. M.D., F.R.C.P.Ed.

Further light is thrown upon the question of whether fish continue to feed in the sea as their genitalia develop, and whether they feed after entering the river, by the study of the differences in the condition of the fish at the mouth and in the upper reaches of the various rivers throughout the season.

Such an investigation also helps to clear up various problems in regard to the migrations of salmon. If there is anything like a general to and fro passage of the fish from the upper waters to the sea and back throughout the season, then the fish at the mouth and fish in the upper reaches will not be sharply marked off from one another. If such migrations do not occur to any extent, then the fish in the upper reaches may be expected to manifest characters different to those in the estuaries.

FEMALE SALMON, 1896.

Table I. gives the length in centimètres, the weight, weight of muscle, and weight of ovaries, both actual and per standard fish, of the female fish examined in grammes :—

[TABLE.

TABLE I.

May and June.

ESTUARY.

No.	River.	Length.	Weight. Actual	Weight. Per Std. Fish.	Weight of Muscle. Actual	Weight of Muscle. Per Std. Fish.	Weight of Ovaries. Actual	Weight of Ovaries. Per Std. Fish.
14	Helmsdale	73	3,610	9,281	2,404	6,180	40	103
15	Spey	73	3,650	9,385	2,340	6,016	42	108
16	Dee	75	3,990	9,456	2,530	5,995	40	95
17	Spey	79	4,295	8,714	2,710	5,496	43	87
18	Dee	71	3,775	10,540	2,370	6,619	52	145
20	Helmsdale	73	3,935	10,120	2,620	6,736	56	144
22	Spey	73	4,100	10,540	2,670	6,864	48	123
23	Dee	67	3,425	11,390	2,174	7,228	16	53
24	Annan	83	6,418	11,220	4,362	7,524	70	122
25	Helmsdale	74	4,095	10,100	2,660	6,568	70	173
27	Dee	73	3,795	9,757	2,444	6,284	57	147
29	Dee	68	3,020	9,605	1,760	5,598	47	150
30	Annan	78	5,915	12,470	3,900	8,219	83	175
35	Annan	70	3,434	10,010	2,280	6,646	71	207

UPPER WATER.

No.	River.	Length.	Weight. Actual	Weight. Per Std. Fish.	Weight of Muscle. Actual	Weight of Muscle. Per Std. Fish.	Weight of Ovaries. Actual	Weight of Ovaries. Per Std. Fish.
11	Spey	69	2,840	8,642	1,860	5,659	34	104
12	Spey	75	3,740	8,863	2,354	5,578	57	135
21	Helmsdale	76	4,055	9,239	2,690	6,128	97	221
28	Helmsdale	73	4,107	10,550	2,700	6,942	114	293
31	Spey	74	3,550	8,762	2,212	5,461	123	301
32	Dee	85	5,447	8,672	3,290	5,358	255	415
33	Dee	73	3,520	9,048	2,234	5,744	144	370

July and August.

ESTUARY.

No.	River.	Length.	Weight. Actual	Weight. Per Std. Fish.	Weight of Muscle. Actual	Weight of Muscle. Per Std. Fish.	Weight of Ovaries. Actual	Weight of Ovaries. Per Std. Fish.
36	Dee	78	4,752	10,010	3,074	6,477	126	266
40	Dee	71	3,810	10,640	2,508	7,004	39	109
41	Dee	70	3,820	11,140	2,210	6,443	150	437
44	Dee	79	5,345	10,840	3,370	6,836	135	274
45	Spey	75	4,225	10,010	2,646	6,270	66	156
46	Helmsdale	71	4,545	12,700	3,024	8,447	73	204
48	Annan	77	5,165	11,310	3,404	7,456	92	202
50	Dee	80	5,170	10,090	3,434	6,707	162	316
51	Spey	74	4,336	10,790	2,840	7,009	80	197
55	Dee	81	5,875	11,050	3,680	6,921	158	297
57	Annan	82	5,300	9,614	3,460	6,276	130	236
58	Spey	79	5,065	10,280	3,090	6,269	236	479
60	Spey	94	9,780	11,780	6,248	7,525	258	311

UPPER WATER.

No.	River.	Length.	Weight. Actual	Weight. Per Std. Fish.	Weight of Muscle. Actual	Weight of Muscle. Per Std. Fish.	Weight of Ovaries. Actual	Weight of Ovaries. Per Std. Fish.
37	Spey	77	4,010	8,782	2,430	5,322	161	353
42	Spey	72	3,504	9,391	2,160	5,789	214	574
43	Helmsdale	66	2,532	8,806	1,532	5,331	103	358
47	Dee	72	3,800	10,180	2,320	6,218	258	691
49	Helmsdale	70	3,490	10,170	2,164	6,310	160	467
52	Helmsdale	73	4,035	10,370	2,470	6,351	241	620

TABLE I.—*Continued.*
October and November.
ESTUARY.

No.	River	Length.	Weight.		Weight of Muscle.		Weight of Ovaries	
			Actual	Per Std. Fish.	Actual	Per Std. Fish.	Actual	Per Std. Fish.
65	Spey	87	8,520	12,940	4,800	7,290	1,025	1,557
*72	Dee	89	7,214	10,230	4,790	6,794	70	99
73	Dee	90	8,184	11,230	4,314	5,920	1,160	1,592
74	Dee	91	8,134	10,800	4,554	6,045	990	1,314
76	Spey	87	6,890	10,470	3,250	4,937	1,270	1,929
77	Spey	81	6,775	12,750	4,030	7,582	425	799
*79	Helmsdale	88	7,415	10,870	5,010	7,350	43	63

* These fish are considered more fully on p. 85, and are not included in Table II.

UPPER WATER.

No.	River	Length.	Weight.		Weight of Muscle.		Weight of Ovaries	
62	Helmsdale	74	3,700	9,133	1,658	4,093	822	2,029
63	Helmsdale	74	3,460	8,541	1,501	3,712	750	1,851
64	Dee	69	3,270	9,956	1,490	4,537	694	2,143
66	Helmsdale	73	3,845	9,886	1,650	4,242	844	2,170
67	Helmsdale	68	2,875	9,145	1,210	3,849	635	2,019
69	Helmsdale	74	4,080	10,070	1,610	3,974	1,100	2,715
70	Dee	66	2,675	9,307	1,140	3,967	590	2,053
78	Spey	84	6,595	11,130	2,700	4,555	1,712	2,888

The analysis of such a Table shows :—

(*a*) *Weight.*

1. The weight per fish of standard length of salmon coming to the mouth of the river increases throughout the season.

TABLE II.

Showing Average Weight per Fish of Standard Length *of Estuary* Salmon (*see footnote*):—

	Spey.	Helms-dale.	Dee.	Annan.	Average.	Limits of Variation.
May and June,	9546	9834	10149	11233	10185	8714—12470
July and August,	10692	12700	10628	10460	10781	9610—12700
Oct. and Nov.,	12053	10 870	11015	—	11648	10270—12940

† Single fish.

The average in table was calculated by adding the weight per standard fish of all the fish of period, and dividing by the number of fish. It is not an average of the rivers.

Mr. Archer, from his analysis of the Berwick-on-Tweed figures (Fishery Reports, Part II., 1895), concludes that fish coming to the mouth of the river gain in weight to the extent of 3 per cent. from May to August. In our series the increase was 5·8 per cent. The statistics at his disposal did not enable him to study the change after August.

2. Salmon in the upper reaches are lighter than those at the mouth, and this difference increases as the season advances, and is nearly twice as great in October and November as earlier in the season.

E

TABLE III.

Showing Average Weight per Fish of Standard Length of Upper-Water Salmon :—

	Spey.	Helms-dale.	Dee.	Average.	Limit of Variation.
May and June,	8756	9895	8960	9139	8642—10550
July and August,	9086	9782	*10180	9616	8782—10370
Oct. and Nov.	*11130	9355	9631	9646	8541—11130

* One fish.

TABLE IV.

Showing Average per Fish of Standard Length of all the Rivers from Tables II. and III :—

	Estuary.	Upper Water.	Difference.
May and June, . .	10185	9139	1046
July and August, . .	10781	9616	1165
Sept. and Oct., . .	11648	9646	2002

(b) Muscle.

3. **Fish** coming to the mouth of the river have, in July and August, a slightly greater amount of muscle, per fish of standard length, than fish in May and June, and a distinctly greater amount than fish **in October and November.**

TABLE V.

Showing Average Weight of Muscle per Fish of Standard Length from Estuaries :—

	Spey.	Dee.	Helms-dale.	Average.	Limits of Variation.
May and June,	6126	6345	6495	6326	5498—7230
July and August,	6768	6732	*8447	6901	6269—8447
Oct. and Nov.,	6603	5982	—	6055	4937—7582

* Single fish.

In July and August there was a rise of 10·8 per cent. in the weight of muscle as compared with May and June ; and a fall of 8·5 per cent. in the October and November fish compared with the mean of the fish from **May to August.**

4. The weight of muscle in fish in the estuaries is greater than in fish in the upper reaches, and this difference becomes more marked as the season advances.

TABLE VI.

Showing Average Weight of Muscle per Fish of Standard Length in Upper Waters:—

	Spey.	Dee.	Helms-dale.	Average.	Limits of Variation.
May and June .	5566	5551	6535	5839	5358—6942
July and August .	5555	6218	5997	5887	5322—6351
Oct. and Nov. .	4555	4252	3974	4116	3712—4555

*Single fish.

If the average of the three rivers be taken the figures are as follows:—

TABLE VII.

Showing Average Weight of Muscle per Fish of Standard Length:—

	Estuary.	Upper Water.	Difference.
May and June .	6326	5839	487
July and August . .	6901	5887	1014
October and November .	6055	4116	1939

Or, comparing the upper-water fish in October and November with the upper-water fish earlier in the year, a loss of nearly 28 per cent. in the weight of the muscle is found.

(c) Ovaries.

5. The weight of the ovaries per fish of standard length increases steadily throughout the season both in the sea and in the river. These results may be compared with those of Hoek and Miescher, and of the Scottish Fishery Board (Fishery Reports 1895, Table p. 30). By these observers the weight of the ovaries is given not as per fish of standard length, but as percentage of the weight of the fish. Hoek's observations on salmon in the Lower Rhine show an increase in the ovaries of from 0·5 per cent. in March to 20 per cent. in November, while Miescher's investigations in the Upper Rhine at Basel give an increase of from 0·75 per cent. in March to 23 in November.

The investigations at the mouth of the Tweed give an increase from 0·75 per cent. in March to 17 per cent. in November:—

[TABLE.

TABLE VIII.

Showing Average Weight of Ovaries per Fish of Standard Length:-

	Spey.		Dee.		Helmsdale.	
	Est'ry	Upper	Est'ry	Upper	Est'ry	Upper
May and June	106	177	118	392	140	257
July and August	286	463	283	691	*204	482
October and November	1428	2890	1454	2083	---	2157

* Single Fish.

Or taking these together :—

TABLE IX.

Showing Average Weight of Ovaries per Fish of Standard Length:—

	Estuary.	Limits of Variation	Upper.	Limits of Variation
May & June	121	53—175	263	104—415
July & Aug.	284	109—479	510	353—691
Oct. & Nov.	1439	1310—1930	2230	1851—2888

This gives an increase from May to November of from 1 to 11·9 in the sea, or 1318 grms. per fish of standard length, and from 1 to 8·5 in the river, or 1967 grms. per fish of standard length. It will thus be seen that although in the upper waters a greater amount of material per fish of standard length is laid on by the ovaries, their *rate of growth*, considering the weight with which they start in May, is quite as great in the sea as in the river.

Throughout every part of the season there is a most marked difference in the ovaries in the upper reaches of the river and at the mouth.

In May and June the ovaries of the fish in the upper reaches are 117 per cent. heavier than those at the mouth, while in October and November they are 55 per cent. heavier.

(d) Length of Salmon.

6. The length of the fish coming to the mouth of the river increases markedly in October and November, while the length of fish in the upper reaches remains fairly constant throughout the season, and corresponds to the length of fish at the mouths from May to August.

TABLE X.

Showing Average Length of Fish :—

	Estuary.	Upper Water.
May and June	74	75
July and August	77	72
October and November	88	73

In drawing conclusions from these figures, it must be remembered that fish of from 8 to 10 lbs. were asked for, but as the supply available was limited, the size of the fish sent was not specially regarded by the senders. Fortunately, the question has been elucidated by the investigations at Berwick-on-Tweed. The following Table shows the results obtained from the examination of female salmon as given in the Fishery Board Report for 1895, pp. 64 to 72.

TABLE XI.

Showing Average Length of Fish in Estuaries:—

	No. of Fish.	Length.
April	29	77
May	43	77·1
June	29	77·3
July	33	77·3
August	30	81
September	18	84
October	37	89
November	45	88·3

FEMALE FISH, 1895.

(a) *Fish from Berwick-on-Tweed.*

In seven of these the ovaries, liver, and muscle—thick and thin—were weighed. The percentage results obtained agreed generally with those obtained from fish caught in the various estuaries during 1896, but as the weight of muscle per fish of standard length was not determined it is unnecessary to give the results of these analyses.

(b) *Fish from Montrose.*

So far as I am aware, all the fish received from Montrose were captured in the lower waters of the North Esk or in Montrose Bay. They are therefore to be considered as estuary fish and compared with the estuary fish of 1896.

Table XII. gives the length, weight, weight of muscle, and weight of ovaries.

A comparison of this Table with Table I. (p. 64) shows a close correspondence, and bears out Conclusions 1 and 4. As regards the growth of the ovaries, it is to be noted that in the North Esk fish in July the ovaries were more developed than in the July and August fish of 1896, but that their average development did not exceed the greatest development observed in the 1896 fish. In the later months the North Esk fish have a somewhat less average size of ovary than the 1896 fish.

[TABLE.

TABLE XII.

FEMALE SALMON, 1895.

No.	Date.	Lgth.	Weight.		Muscle.		Ovaries.		Remarks.
			Actual	Per Fish of Stnrd. Length.	Actual	Per Fish of Std. Lgth.	Actual	Pr Fsh of Std. Lgth.	
10	23.7	76	4130	9410	2590	5899	218	496	North Esk.
11	,,	61	2055	9053	1262	5560	725	319	Grls, Montrose
12	,,	78	5390	11359	3630	7671	218	459	Montrose Bay.
Average · ·				9940		6377		424	
24	3.9	81	5954	11235			278	523	North Esk.
25	,,	90	7704	10568			240	331	,,
Average · ·				10901				427	
29	26.10	76	4555	10376	2680	6105	317	722	,,
31	28.10	94	9300	11197	5560	6694	995	1199	,,
39	12.12	91	7636	10293	3740	4322	1588	2108	,,
Average · ·				10592		5707		1343	

MALE SALMON.

For the study of male fish the amount of material at our disposal was
unfortunately very limited.

TABLE XIII.

*Showing the Length, Weight of Muscle, and Weight of Testes, both Actual
and per Fish of Standard Length :—*

May and June

ESTUARIES.

No.	River.	Length.	Weight.		Muscle.		Testes.	
			Total.	Per Std. Fish.	Total.	Per Std. Fish.	Total.	Per Std. Fish.
13	Spey . .	77	4090	8958	2670	5850	5	10·97
19	Annan . .	80	4435	8662	2870	5605	8	15·62
	Averages .	78·5	4262	8810	2770	5727	6.5	13·3

UPPER WATER.

26	Helmsdale	83	5557	9737	3544	6209	24	42·05
34	Dee . . .	68	2650	8429	1654	5261	14	44·46
		75·5	4103	9083	2599	5735	19	43·2

July and August.

ESTUARIES.

53	Annan . .	77	5290	11588	3564	7800	14	32·86
56	Spey . .	74	4480	11056	2810	6934	8	19·74
59	Spey . .	87	7010	10644	4604	6991	24	36·44
61	Annan . .	84	5650	9583	3230	5450	45	75·92
	Averages .	80·5	5607	10705	3552	6794	23	41·2

TABLE XIII.— *Continued.*

July and August.

UPPER WATER.

No.	River.	Length.	Weight.		Muscle.		Testes.	
			Total.	Per Std. Fish.	Total.	Per Std. Fish.	Total.	Per Std. Fish.
38	Helmsdale	74	3870	9551	2454	5957	10	44·42
39	Dee . . .	77	3885	8510	2420	5301	44	96·38
54	Dee . . .	79	3880	7810	2334	4694	101	204·83
		77	3878	8623	2386	5317	52	115·2

October and November.

ESTUARIES.

No.	River.	Length.	Weight.		Muscle.		Testes.	
71	Dee . . .	108	12,760	10129	7610	6041	270	214·33
75	Dee . . .	68	2890	9192	1660	5280	96	305·34
	Averages .	88	7285	9660	4635	5660	183	260

UPPER WATER.

No.	River.	Length.	Total	Per Std	Total	Per Std	Total	Per Std
68	Dee . . .	74	3280	8094	1712	4225	109	269·

(a) *Weight.*

1. The weight per fish of standard length increases in the salmon coming to the mouths of the rivers to August.

TABLE XIV.

Showing Average Weight per Fish of Standard Length in Estuaries :

	Weight per Fish of Std. Length.	Limits of Variation.
May and June, · · ·	8810	8662— 8958
July and August, · · ·	10705	9535—11588
October and November, · ·	9660	9192—10129

2. The difference between the weight of the fish at the mouth and in the upper waters of the rivers is practically nil in May and June, but as the season advances the weight of the fish in the upper reaches becomes less than that of fish in the lower reaches.

In May and June there is a difference of only 3 per cent. In July and August the estuary fish are 19 per cent heavier than the upper-water fish, and in October and November 16 per cent. heavier.

TABLE XV.

Showing Average Weight per Fish of Standard Length in Upper Waters:—

	Weight per Fish of Standard Length.	Limits of Variation.
May and June, - -	9083	8430—9737
July and August, - -	8623	7810—9551
October and November, -	8094	8094

(b) *Muscle.*

The weight of muscle in fish coming to the estuaries is greatest in July and August and least in October and November. In May and June the weight of muscle is the same in fish in the upper waters as in fish in the estuaries, but in July and August it is 22 per cent. less and in October and November 25 per cent. less.

TABLE XVI.

Showing Average Weight of Muscle per Fish of Standard Length:—

	Estuaries.	Upper Waters.
May and June, - -	5727	5735
July and August, - -	6794	5317
October and November, -	5660	4225

(c) *Testes.*

The weight of the testes increases in fish in the estuary and in the upper waters throughout the season, and the increase is as marked in fish in the estuaries as in fish in the upper waters.

TABLE XVII.

Showing Average Weight of Testes per Fish of Standard Length:—

	Estuaries.	Upper Waters.
May and June, - -	13·3	43·2
July and August, - -	41·2	115·2
October and November, -	260·0	269·0

KELTS.

It is a matter of no little interest to trace the changes which take place in the salmon between the time of spawning and their return to the sea.

It is supposed by some that kelts feed voraciously. Of the two salmon captured at Basel in which Miescher Ruesch found traces of food, both were kelts.

The evidence adduced by Dr. Gulland shows that the lining membrane of the stomach is regenerated in the kelt stage, while Dr. Gillespie's observations show that the digestive power is greater in kelts than in unspawned salmon. In the kelts examined by us food was never found in the stomach, but, in nearly all, the gall-bladder contained a greater or less quantity of bile.

Miescher Ruesch describes the changes in the salmon after spawning as follows : —

After a vivid description of the characters of the fish on the spawning beds, he says (p. 215) :

" How altogether different is the picture if we have the opportunity to see the animal ten days, or, better, two weeks, after spawning. The skin is again blueish, shining and clear, the ulcers cicatrized and healing, the flesh transparent and free of oil granules. The heart fibres also participate in the regenerative change; in the intestine is no trace of food. On the other hand, the ovary contains sometimes more, sometimes fewer eggs, which are embedded in a serous or somewhat purulent effusion of the follicular membrane, and are evidently shrinking and being absorbed. They are thus a sort of nourishment, a provision (Zehrgeld) for the return journey. But I ascribe the chief importance to the pale, shrunken, and folded follicular membrane. The collateral vessels of the ovary are closed through vascular contraction. The salmon is like a patient who has had a leg amputated after the application of an Esmarch's bandage. Its blood courses in a narrow circulation, thus with higher pressure, and supplies a less amount of oxygen-requiring matter than formerly. The circulation is again sufficient for its task, and the trunk muscles becomes normal. . . . . The little nutrient matter coming from the ovary greatly helps the reconvalescence of the muscle."

In the spring of 1897, 22 kelts were received from the mouth of the Spey between the 17th March and the 21st May, and other three kelts were procured in 1895-96.

The weight of muscle, ovaries, etc., was determined in eleven of these, and a detailed analysis was made of four.

Table XVIII. gives the length, weight, weight per fish of standard length, weight of musculature total and per fish of standard length, and weight of ovaries total and per fish of standard length. The length is in centimètres and the weight in grammes :—

[TABLE.

TABLE XVIII.

KELTS.

No.	Date.	Length.	Weight.		Muscle.		Ovaries.		
			Total	Per Std. Fish.	Total	Per Std. Fish.	Total	Per Std. Fish.	
14	1895. July 23	83	3,995	6,986	2,400	4,196	42·5	74	N. Esk.
8	1896. March 17	66	2,458		1,512	—	11	—	N. Esk.
9	,, 17	65	2,280		1,222	—	14	—	N. Esk.
80	1897. March 17	87	5,434	8,243	3,070	4,658	29	44	Spey.
81	,, 18	93	6,847	8,518	4,000	4,974	52	64	—
82	,, 18	92	6,235	8,004	3,500	4,492	65	83	—
83	,, 18	89	4,800	6,866	2,489	3,548	42	60	—
84	April 1	93	6,485	8,064	3,916	4,807	38	47	—
85	,, 5	94	6,455	7,771	3,610	4,346	40	48	—
86	,, 8	94	6,895	8,252	3,730	4,490	54	65	—
87	,, 15	91	6,280	8,389	—	—	—	—	—
88	,, 15	96	6,110	7,009	—	—	—	—	—
89	,, 15	65	1,877	6,829	—	—	—	—	—
90	,, 21	91	5,975	7,934	—	—	—	—	—
91	,, 21	89	5,510	7,882	—	—	—	—	—
92	,, 27	91	5,525	7,311	3,120	4,140	31	41	—
93	,, 30	98	6,955	7,389	—	—	—	—	—
94	May 5	92	6,080	7,869	—	—	—	—	—
95	,, 5	90	5,735	7,867	—	—	—	—	—
96	,, 5	67	2,365	7,865	—	—	—	—	—
97	,, 6	85	5,060	8,289	—	—	—	—	—
98	,, 10	96	5,085	6,975	—	—	—	—	—
99	,, 12	69	2,282	6,943	—	—	—	—	—
100	,, 14	91	6,035	8,008	—	—	—	—	—
101	,, 21	85	4,867	7,925	—	—	—	—	—
	Average	83·8	5091	7,755	2,980	4,487	38	56	—

From the length of the fish it is manifest that most of these are kelts of the late-coming fish—October and November salmon.

Whether **the smaller fish** which come from the sea earlier in the year descend **at an earlier date is** not shown by these results.

The table shows :—

(1) That, as might be expected, the kelts are lighter per fish of standard length **than the** unspawned fish.

(2) **The** amount of muscle is rather greater than in the unspawned fish. This is of interest in connection with Miescher Ruesch's description of the possible regeneration of the muscle from the ova retained **in** the peritoneal cavity.

(3) The weight of the ovaries is smaller than in the fish coming to the rivers **in** May and June. It should be mentioned that in all these kelts a considerable number of unshed ripe **ova were** always found in **the abdominal cavity.**

As compared with winter salmon 72 and 79 (Table I., p. 65), the weight of ovaries in the kelt is **not** appreciably smaller. The average weight of ovaries in 72 and 79 was 81 grms. per fish of standard length. **The** average **weight in the** kelts was 58 grms. ; but while the ovaries of 79 were only 66 grms. per fish of standard length, kelt 82 had ovaries of 83 grms.

GENERAL CONCLUSIONS FROM THESE RESULTS.

These figures throw important light upon several questions :—

1st. They **confirm** the conclusions arrived at by Mr. Archer in his

Report of 1895, that fish continue to feed in the sea at least till the end
of August. The marked diminution in the amount of muscle in fish
reaching the estuaries in October and November would seem to show
that the supply of food is insufficient to yield the material necessary
for the rapidly growing genital glands, and that therefore the solids of
the muscle have to be drawn upon, or, at least, that accumulation of material
in the muscles is prevented. The steady increase in weight per fish of
standard length throughout the season seems to indicate that they continue to feed even after August and September, though, as will be
shown later (p. 86), the flesh contains about 5 per cent. more water
in October and November than in July and August, while the increase
of weight is only 3·7 per cent.

2nd. The fall in the amount of muscle from the early to the late part
of the season in fish in the upper reaches supports the conclusions
arrived at by Meischer Ruesch, and by Drs. Gulland and Gillespie,
that the salmon does not feed in fresh water.

3rd. Light is thrown on the question of whether fish entering the
river early in the year go straight up and occupy the upper reaches,
leaving the lower parts to be occupied by the later-coming fish.

The following facts bear specially upon this :—

(1) The length of the fish coming to the mouth of the rivers is practically the same from May to August (pp. 68 and 69). But in October and
November a markedly larger class of fish appears in the estuaries. In
the upper reaches, however, the size of the fish remains constant till
October. This would seem to show that from early spring to August
the fish press upwards, but that after this the later arrivals occupy the
lower reaches of the river.

(2) The fact that the weight per fish of standard length steadily rises
in the estuaries throughout the season, and rises in the upper waters also
until July and August, seems to show the passage of fish to the upper
reaches during these months. But there is no increase of the weight of
the upper-water fish in October and November corresponding to the
increase in weight of those at the estuaries. This supports the view that
the early-coming fish pass on to the upper reaches.

(3) The weight of the musculature increases in fish coming to the
mouth of the rivers from May to August. There is also a slighter
increase in fish in the upper reaches, due probably to the immigration
of these more muscular fish from the mouth. But in October and
November, while the estuary fish show only a slightly less developed
musculature, the fish in the upper reaches show a very marked diminution, indicating that immigration from below has practically stopped.

These conclusions are further supported by the evidence adduced in a
footnote of the Appendices of the Twelfth Annual Report to the Fishery
Board, 1893, pp. 55 and 56 :—

" If fish ascended the rivers in the spring of the year only to rid themselves of sea-lice, as some consider, they might be expected to ascend
only a short distance ; whereas, it seems, on the contrary, that in some
rivers, at any rate, they press up immediately to the head waters.
Thus, in the Forth District, it is said to be the floods in January and
February that induce the fish to ascend to Lochs Vennacher and Achray.
In the Tay District, the fishing is best in the Loch Tay in the early
months of the year ; and I am informed by a gentleman well acquainted
with the fishings at the foot of the falls of Tummel, that it is the March
floods that give a successful season in those waters, whereas the autumn
fish do not ascend so far. The superintendent of the river Dee, in

Aberdeenshire, informs me that the spring fish in that river press up stream into the tributaries, some 15 to 20 miles above Braemar ; whereas the autumn fish, which collect in the greatest numbers between Banchory and Ballater, are seldom found above Braemar. He distinguishes between spring and autumn fish from the fact that fish taken ascending the river in the spring are, as a rule, small fish, whereas those taken in the autumn are considerably larger. Small fish, corresponding in size to the spring fish and much discoloured, showing they had been in the fresh water a long time, are found in the upper part of the river, and are the first to spawn ; whereas the fish in the lower part of the river resemble, in size, those caught during the autumn, and present the appearance of having more recently left the sea.

" In the district of the river Ness, the fish run straight through the river Ness and Loch Ness into the river Garry. The river Garry flows into the upper end of Loch Ness. Indeed, it is said that, although the Oick and Garry fish must pass up the Ness, scarcely any settle there, no fish being taken in the Ness by the fly before July. The same is said of the Orchy on the West Coast. The Orchy flows into Loch Awe, which is connected with the sea by the river Awe. The Awe is a late river, the heaviest salmon being got in the autumn, while the Orchy below the falls is best in the spring. The spring fish are said, as a rule, to come right up to the long, deep, rocky pool below the falls. In the river Severn in England, Mr. Willis Bund, in his book *Salmon Problems,* says that there is a spring run of small salmon, weighing from 8lbs. to 15lbs., in February and March. These are very strong active fish, and press up the river at once, those getting to the top forming the early spawners. These instances are sufficient to show that in some rivers, at any rate, clean fish, having their roe very slightly developed, ascend at once to the head waters of the river.

" It will be observed, also, that the observations of Mr. Willis Bund on the Severn, and the superintendent of the water bailiffs on the Dee, tend to show that the spring fish are the early spawners, and form the breeding stock in the head waters of the rivers. The observations of Mr. Ffennel, one of the commissioners appointed under the Salmon Fisheries (Scotland) Act of 1862, correspond with those of Mr. Willis Bund and the superintendent of the Dee. In his evidence before the Select Committee of the House of Lords in 1860, he said that the salmon which entered the rivers in November, December, January, February, and March, spawned in the following October. He said that in the river he had lived upon, the Suir, he had watched salmon from childhood ; in February new fish came up plentifully, the water got very low and clear, and they could see them in the pools through the summer. He had, therefore, no doubt that salmon would live and thrive for a whole twelve months in fresh water. He further stated that these fish, although discoloured, remained very fat and exceedingly good to eat up till midsummer.

" At Sand in Norway, where I watched the habits of salmon for some years, the early fish having high crests and slight development of roe, came into the river in June. These fish, as soon as they could pass the first fall, ran right up to the second, whereas the lower pools did not contain fish in any numbers until late on in the autumn."

4th. It has been sometimes suggested that the fish which ascend the rivers early in spring do not remain there throughout the season, but that they may again descend to the sea to feed, and again ascend the river to spawn later in the season. That salmon may and do move up and down the rivers during the summer and autumn there can be no doubt, but, so far as I am aware, there is absolutely no evidence that once having fairly entered the river they ever return to the sea in any considerable number. Were this the case, did a constant to-and-fro migration from sea to river and river to sea occur, we should not expect to find any marked difference between the fish in the estuaries and the fish in the upper reaches. But our figures show that as regards weight, as regards weight of muscle, and as regards weight of ovaries and testes, there is a marked difference, becoming more accentuated as the season advances, between fish in estuary and fish in the upper waters. And the more carefully these figures are analysed the more forcibly do they tell against such a to-and-fro movement.

In considering the total weight of the salmon, it might be thought that the very marked overlap between the heaviest fish in the upper reaches and the lightest fish in the estuary favoured this view.

Weight per Standard Fish.

	Estuary. Lightest.	Upper Reaches. Heaviest.	Overlap.
May and June	8,714	10,055	1,341
July and Aug.	9,610	10,370	760
Oct. and Nov.	10,270	11,130	360

Analyses of the weights of the musculature and of the ovaries shows that no significance can be given to these figures.

In the first place the very small overlap between the smallest ovaries of the upper-water fish and the largest ovaries of the estuary fish opposes the idea of there being any free exchange between these situations.

Weight of Ovaries per Standard Fish.

	Estuary. Heaviest.	Upper Reaches. Lightest.	Overlap.
May and June	183	175	72
July and Aug.	353	479	126
Oct. and Nov.	1,930	1,351	None.

But a comparison of the musculature is the most convincing. In May and June while the fish are running, when no time has been allowed for any difference to develop between the fish leaving the sea and those in the upper reaches, the overlap between the largest musculature of the upper-water fish and the smallest musculature of the estuary fish is fairly marked. By July and August the overlap has practically ceased to exist, and by October and November the heaviest musculature of the upper-water fish is less than the smallest musculature of the estuary fish.

Weight of Muscle per Standard Fish.

	Estuary. Smallest.	Upper Reaches. Largest.	Overlap.
May and June - -	5,498	6,940	+ 1,442
July and Aug. - -	6,269	6,350	+ 81
Oct. and Nov. - -	4,937	4,550	None

Such conditions would be impossible were there a free to-and-fro migration.

5th. The close correspondence between the fish from the Helmsdale, Spey, and Dee appears to indicate that these may be regarded practically as one river. The experiments on marking fish at present being conducted by Mr. Archer, so far as they go, support this idea.

CHEMICAL CHANGES IN THE SALMON IN FRESH WATER.

PRELIMINARY CONSIDERATIONS.

By D. NOËL PATON, M.D., F.R.C.P. Ed.

It has been already shown that the salmon leaving its marine feeding ground ascends the river and remains until the spawning time, often for several months, without any supply of nourishment from without.

During this period it must thus subsist upon the store of material in its body brought from the sea.

Hence a comparison of the condition of the salmon in the upper reaches and in the estuary of the river at different periods will make manifest the progress of the changes which go on in the fish during its prolonged fast, and will explain how the material for the growth of the genitalia and for the muscular energy required in the ascent of the stream is obtained.

The following sections deal with these matters :—

The question has been already investigated and discussed by Miescher Ruesch (*loc. cit.*), and before proceeding to give the results of the present enquiry it is necessary to carefully consider his important work.

1. His investigations were practically confined to fish obtained in the upper waters of a long river—at the Basel-Laufenberg fisheries, about 500 miles from the mouth of the Rhine.

2. His observations extended over three seasons, 1877 to 1880. From the measurement and weighing of 470 salmon of from 840 to 917cm. in length from May to November, he concludes that on an average the fish loses about 6 per cent. of its weight from May 22nd to October 12th.

3. From observations on 39 salmon he states that during this period the ovaries increase from 1·1 per cent. of the body weight to about 23 per cent.

4. Analyses of the muscle in 15 salmon show that in the earlier part of the year—in July and August—there is about 17·5 per cent. of albumin (proteids), while later—in November and December—there is only 13 per cent. He concludes that this loss of albumin alone from the lateral trunk muscles is sufficient to cover four-fifths of the weight of the full-grown ovaries.

5. From the difference between the weight of the total solids and the albumin he calculates the amount of fat in the muscle, and shows that in August it is about 5 per cent. and in November about 2·2 per cent.

6. To study the changes of material more accurately two salmon were taken, one on the 7th and one on the 4th of August 1878, one an

average fish, the other an exceptionally thin fish, and in these fish the various organs were carefully weighed and analysed.

From his previous average tables he calculates what the weight of muscle, ovaries, etc., of these fish would have been on November 1st, and what would have been the proportion of albumin and fats in these organs. Then balancing the condition in August with the supposed condition in November he concludes that the amount of albumin and fat lost from the muscle is more than sufficient to build up the ripe ovaries.

A number of other points of interest are also considered in the enquiry, such as the transference of phosphorus from muscle to ovaries ; the nature of the phosphorus compounds in each, and the nature of the ovarian fluid. Some of these points will have to be dealt with in a later part of this Report, some of them are of merely incidental interest and need not be considered.

A careful consideration of Miescher's work shows that it leaves untouched several questions of interest, and that the conclusions arrived at are hardly warranted by the evidence produced.

In the first place it leaves unconsidered the changes which the fish undergo in passing from the sea up the river, and, therefore, does not deal with the question of how much of the stored material in the fish goes to the formation of the genitalia and how much is used as a source of energy.

In the second place, in order to arrive at his conclusions, he assumes that from August to November no fresh fish are coming to the Basel water from the sea, since if this were the case his average results could not be applied to the fish analysed. He argues at length against the possibility of such an immigration of lower-water fish on the ground that, if they came, there must be two classes of fish in the upper waters, one poor in material, consisting of the fish which came up in spring, and one richer in material, consisting of the fish which continued to feed in the sea and came up with their genitalia more developed and with a greater supply of nourishment in the muscle. According to Miescher such two classes do not exist. It is, however, possible that the wide divergence in the weights of individual fish noted by him—as much as 20 or 30 per cent.—may be explained by the arrival of these better nourished fish. The observations recorded on p. 75 of this paper show pretty clearly that in the shorter Scottish rivers, fish leaving the sea in August do pass to the upper reaches. Again, in this argument he makes the assumption, which he does not attempt to prove, that fish continue to feed in the sea all through the autumn.

The conclusions which Miescher bases upon his analyses are so far-reaching and important that they must be considered with critical care. It seems hardly justifiable to base such conclusions upon so limited a number of observations. Still less is it safe to assume that any two particular cases will accord strictly with the average results. It must be remembered that the detailed examination of only two fish is recorded.

So far as the analyses are concerned it may safely be concluded that in the hands of so able a chemist these were properly carried out. Two points, however, require to be pointed out. First, that the fats were not directly determined, but that the difference between the total solids and the albumin was taken to represent the fats. Second, for the analyses of the muscle he says (p. 179)—" An einer Reihe von Juli und Augustsalmen wurde desshalb, immer von desselben Stelle genommen, der Eiweissgehalt des grossen Seitenrumpfmuskels bestimmt." Apparently no cognisance was taken of the marked difference between the trunk muscle generally—the " thick " of the fish and the belly muscle or "thin." From an analysis of one piece of muscle the composition of the whole was calculated. He gives no analyses of the ovaries, and says that he does not possess a sufficient number of completed analyses of the albumin of these structures, and he adopts 30 per cent. as probably too high an amount.

It was for these reasons that a re-investigation and amplification of Miescher's work appeared desirable, especially as his results have been generally accepted and have been made the basis of very general conclusions in regard to the metabolic processes in starvation.

In approaching the subject we had the advantage of a supply of fish from the mouth and upper waters of the rivers at the same periods, and whole fish were placed at our disposal, so that the weight of the muscle, ovaries and other structures could be directly determined in each case, and complete analyses undertaken.

METHOD OF COMPARISON.

To compare the salmon of different sizes analysed during the enquiry, all weighings of organs and of constituents of organs were expressed in terms of fish of standard length—100 cm. (See page 6.)

It has been already shown that there is no essential difference between the salmon from the Helmsdale, Spey, and Dee, and that at all seasons they are practically in the same condition. Hence salmon from the different rivers were considered as comparable with one another.

Instead of following Miescher's method of determining the nature of the exchanges between muscle and ovaries, the following procedure was adopted :—

An average of the weighings and analyses of the different organs of the fish from the estuaries and from the upper reaches of the rivers in the three periods—

May and June,
July and August,
October and November

were prepared, and from the differences between these, conclusions were drawn as to the nature and extent of the change.

Since it is impossible to know whether the fish captured in the upper waters in May and June had left the sea during these months, and had not, in part at least, come up earlier in the year, no definite conclusions are drawn from a comparison of the fish from these two situations during this period.

On page 75 clear evidence has been advanced that the fish which run in July and August pass up and mingle with the earlier run fish. It therefore seemed justifiable to compare the upper-water fish of July and August with the fish taken in the estuaries from May to August.

It has also been shown (page 75) that the fish which leave the sea in October and November already full of more or less ripe spawn do not immediately pass up to the upper reaches, but probably remain and spawn in the lower part of the river. Hence the upper-water fish of October and November have to be compared with the fish leaving the sea in the earlier part of the year, from May to August, or with the fish already in the upper water in July and August.

The following table shows this system of comparison :—

Upper Water.   Lower Water.
July and August Fish compared with May to August Fish.
October and November Fish „ „ May to August Fish.

F

SALMON ANALYSED.

Of the 55 female fish received during 1896, 36 were analysed. These are grouped as follows : --

TABLE I.

	ESTUARY FISH.		UPPER-WATER FISH.	
	Fish Received.	Fish Analysed.	Fish Received.	Fish Analysed.
May and June,	14	7	7	6
July and August,	13	6	6	5
October & November,	7	5	8	7

Classified according to the rivers from which they were received : —

TABLE II.

		Helmsdale.	Spey.	Dee.
May and June,	Mouth,	2	2	3
	Upper,	2	3	1
July and August,	Mouth,	1	2	3
	Upper,	3	2	1
October and	Mouth,	1	1	3
November,	Upper,	5	0	2

Seven salmon received from Montrose in 1895 were also analysed. Of the 14 male fish received in 1896, 6 from the estuaries and 6 from the upper reaches of the rivers were analysed.

7.—THE CHANGES IN SOLIDS AND WATER OF MUSCLES, AND GENITALIA OF THE SALMON IN FRESH WATER.

By D. NOEL PATON, M.D., F.R.C.P.Ed.

In investigating the chemical changes in the tissues throughout the season, we may begin by the consideration of the total solids of the muscles and genitalia in fish from different stations throughout the season.

Method.

The total solids were determined as follows :—

The fat was extracted from the tissue in the manner afterwards to be considered (p. 93). The residue was then dried at 110° C. and weighed. To this the weight of the fat was added to give the total solids.

Female Salmon, 1896.

Table 1. gives the per cent. of solids in the thick and thin of muscle, the solids calculated per fish of standard length, the per cent. of solids and the solids per fish of standard length in the ovaries.

TABLE I.
ESTUARY FISH.
I.—May and June.

No.	River.			Muscle.			Ovaries.	
				Per Cent.		Total per Fish of Standard Length.	Per Cent.	Total per Fish of Standard Length.
				Thick.	Thin.			
16	Dee .	.	.	34·3	38·7	2110	33·3	31
20	Helmsdale	.	.	33·7	38·3	2310	36·7	52
25	Helmsdale	.	.	31·2	35·9	2130	34·5	59
27	Dee .	.	.	33·9	42·1	2250	35·6	52
		Average,		34·2	38·7	2210	35·0	47·3
15	Spey .	.	.	27·4	29·1	1660	33·7	36
17	Spey .	.	.	31·7	33·9	1760	30·8	26
29	Dee .	.	.	29·4	35·6	1720	27·8	11
		Average,		29·5	32·8	1710	29·7	31
	Average of	two sets,		31·6	36·2	1990	32·9	42·4

TABLE I.—*Continued.*

II.—July and August.

No.	River.		MUSCLE.			OVARIES.	
			Per Cent.		Total per Fish of Standard Length.	Per Cent.	Total per Fish of Standard Length.
			Thick.	Thin.			
36	Dee .		33·0	39·0	2230	36·2	96
40	Dee .	.	32·1	37·0	2320	30·6	33
45	Spey .		33·9	37·2	2170	32·0	49
51	Spey .	.	32·5	34·3	2600	34·0	66
55	Dee .		33·3	37·1	2960	39·2	116
		Average,	32·9	36·1	2270	34·4	72

III.—October and November.

No.	River.		MUSCLE.			OVARIES.	
65	Spey .	.	27·7	29·2	2040	38·1	593
73	Dee .		27·8	35·7	1750	38·9	618
74	Dee .		22·9	29·2	1470	39·7	514
		Average.	26·1	31·3	1750	38·7	545
72	Dee .		35·9	34·4	2414	28·7	28·4
79	Helmsdale		35·2	33·1	2549	28·4	17·9
		Average,	35·5	33·7	2481	28·5	23·1

UPPER-WATER FISH.

I.—May and June.

TABLE I.— *Continued.*

II.—July and August.

No.	River.	MUSCLE.			OVARIES.	
		Per Cent.		Total per Fish of Standard Length.	Per Cent.	Total per Fish of Standard Length
		Thick.	Thin.			
37	Spey	31·2	35·0	1700	38·6	136
42	Spey	28·3	30·8	1670	39·3	225
43	Helmsdale	28·7	30·8	1540	37·7	131
49	Helmsdale	28·5	33·1	1860	38·3	178
	Average,	29·2	32·4	1690	38·5	168

III.—October and November.

No.	River.	MUSCLE.			OVARIES.	
62	Helmsdale	20·9	20·5	850	37·6	763
63	Helmsdale	21·6	22·1	805	38·6	714
64	Dee	21·98	25·1	1030	38·7	816
66	Helmsdale	22·5	28·5	1010	39·2	850
67	Helmsdale	20·6	22·5	811	39·7	801
69	Helmsdale	21·5	22·6	863	34·7	940
70	Dee	19·4	22·2	797	35·5	727
	Average,	21·2	23·3	880	37·7	801

In these Tables certain fish require special consideration. In the estuary fish of May and June, Nos. 15, 17, and 29 were found to differ so markedly from all the other May to August fish, that they were excluded from the general class. Nothing special is noted as to the appearance of No. 15. In No. 17 the gall-bladder was found to be empty, and in 29 there were no sea lice, and some old wounds were present on the left side. In drawing conclusions from the table it seemed better to exclude these fish, though their inclusion does not invalidate any of the conclusions arrived at.

In the estuary fish of October and November, Nos. 72 and 79 differ from the others in the large amount of solids in the muscles and the small development of the ovary. They are, in fact, examples of the "winter salmon" of Miescher Ruesch, fish not going to spawn till the ensuing season. They are, therefore, excluded from the general class.

In the upper-water fish of May and June, one, No. 11, appears abnormal, and it has been excluded in drawing conclusions. Its inclusion would merely emphasise the distinction between the upper and lower water fish.

On taking the average results of this table we find the following :—

[TABLE.

TABLE II.

	SOLIDS OF MUSCLE.					
	Estuary.			Upper Reaches.		
	Per Cent. of Solids.		Solids per Fish of Standard Length.	Per Cent of Solids.		Solids per Fish of Standard Length.
	Thick.	Thin.		Thick.	Thin.	
May and June,	33·2	38·7	2240	29·8	32·7	1710
July and August, .	32·9	36·1	2270	29·2	32·4	1690
October and November,	26·4	31·3	1750	21·2	25·3	880

	SOLIDS OF OVARIES.			
	Estuary.		Upper Reaches.	
	Per Cent. of Solids.	Solids per Fish of Standard Length.	Per Cent. of Solids.	Solids per Fish of Standard Length,
May and June,	35·0	47·3	36·3	95·5
July and August, . .	34·4	72·0	38·5	168·0
October and November,	38·7	545·0	37·7	801·0

Such a table shows : — As regards

(*a*) MUSCLE.

1. The percentage of solids is throughout the season markedly higher in the fish at the mouth than in the fish in the upper reaches of the river.

2. The percentage remains unaltered both at the mouth and in the upper reaches till August.

3. In October and November the percentage of solids falls markedly both at the mouth and in the upper reaches.

4. The October and November salmon in the estuary contain in the thick 7 per cent. and in the thin 8 per cent. more water than the salmon in May, June, July, and August, and thus, though the weight of the muscle does not show a diminution, there is actually a marked diminution in its solid constituents, and its weight is kept up by the addition of water. In the October and November fish in the upper waters, the thick contains 12 per cent. and the thin 15 per cent. more water than the thick and thin of fish in the estuary earlier in the year. This development of a more watery condition of the flesh is of considerable importance in estimating the food value of the salmon.

5. The solids in the fish of standard length undergo no alteration in the estuary fish from May to August, but in October and November there is a marked decrease.

In the upper-water fish there is practically no change till August, but there is a very marked diminution in October and November.

(b) Ovaries.

1. From May to August the percentage of solids is higher in the fish from the upper reaches than in those from the mouth of the rivers. In October and November the ovaries of fish at the mouth have as high a percentage of solids as those in the upper waters.

2. Throughout the season the total solids of the ovaries in the fish of standard length are very markedly less in fish at the mouth than in fish in the upper waters.

From these tables a balance of Loss of Muscle and Gain of Ovaries may be struck :—

A. Balance to August.

Average solids of muscle per fish of standard length in estuary fish from May to August, - - -	2240
Average solids in muscle per fish of standard length of upper-water fish in July and August, -	1690
Loss. -	550
Average solids per fish of standard length in ovaries of estuary fish from May to August, - - -	59
Average solids in ovaries in upper-water fish in July and August, - - - - - -	168
Gain. -	109

If 109 of muscle solids go to ovaries, then 441 grms. will be available as a source of energy.

B. Balance to November.

Solids in muscle of estuary fish, May to August, -	2240
Solids in muscle of upper-water fish, October and November, - - - - - -	880
Loss. -	1360
Solids in ovaries of estuary fish, May to August, -	59
Solids in ovaries of upper-water fish, October and November, - - - - - -	800
Gain, -	741

If 741 of solids of muscle go to ovaries, 619 will be available as a source of energy.

The amount of solids lost by the muscle is not only amply sufficient to yield the solids gained by the ovary, but a large surplus is left.

FEMALE SALMON—1895.

Table III. gives the results of the analyses of the amount of solids in the North Esk fish during 1895.

It corresponds closely to Table I., except as to the solids of the ovaries. Here in July both the percentage amount and the amount per fish of standard length are markedly higher than in the corresponding fish of 1896.

TABLE III.

SOLIDS.

No.	Date.	Muscle.			Ovaries.	
		Per Cent.		Total per Standard Fish.	Per Cent.	Total per Standard Fish.
		Thick.	Thin.			
10	23.7	33·8	34·7	2004	11·3	204
12	..	35·4	36·6	2737	10·2	175
Average, ..		34·6	35·6	2370	10·7	389
29	26.10	27·5	30·2	1722	28·0	274
31	28.10	31·6	37·9	2320	35·5	161
39	12.12	23·2	19·9	1000		
Average, ..		27·4	29·3	1614	38·0	367

MALE FISH.

Though the testes increase in size about twenty fold, they do not attain anything like the same proportionate weight of the fish as do the ovaries. In a fish of standard length in the male, 257 grms. of material are laid on to the testes; in the female, 2150 grms. are gained by the ovaries. It is, therefore, highly probable that since the muscles in the female contain sufficient material for the growth of the ovaries, in the male the stored material should be amply sufficient for the growth of the testes. Unfortunately the number of male salmon at our disposal was small.

Table IV. gives the results of our investigations on the changes of the solids in the muscle and testes :—

[TABLE.

TABLE IV.—SOLIDS.

Estuary.

MAY AND JUNE.

No.	River.		Muscle.			Testes.	
			Per Cent. Thick.	Per Cent. Thin.	Total per Standard Fish.	Per Cent.	Total per Standard Fish.
13	Spey		29·2	34·5	1781	19·8	2·14
19	Annan . . .		31·9	35·6	1873	20·2	3·30
		Average,	30·5	35·0	1827	20·0	2·72

JULY AND AUGUST.

No.	River.		Thick	Thin	Total	Per Cent	Total
56	Spey		24·1	38·1	2379	17·1	3·38
59	Spey . . .		35·0	39·6	2526	18·1	6·09
		Average,	29·5	38·8	2452	17·6	4·73

OCTOBER AND NOVEMBER.

No.	River.		Thick	Thin	Total	Per Cent	Total
71	Dee		25·9	28·8	1607	26·3	56·1
75	Dee . .		24·9	26·2	1333	25·0	76·1
		Average,	25·4	27·0	1470	25·6	66·0

Upper Water.

MAY AND JUNE.

No.	River.		Thick	Thin	Total	Per Cent	Total
26	Helmsdale		32·5	35·5	2065	19·3	8·1
34	Dee		28·4	30·0	1484	18·1	8·0
		Average,	30·4	32·7	1774	18·7	8·0

JULY AND AUGUST.

No.	River.		Thick	Thin	Total	Per Cent	Total
38	Helmsdale		30·5	32·1	1842	17·4	7·7
39	Dee		30·0	34·6	1648	14·8	14·3
54	Dee		27·2	28·1	1285	16·8	33·9
		Average,	29·6	31·6	1592	16·3	18·6

OCTOBER AND NOVEMBER.

No.	River.		Thick	Thin	Total	Per Cent	Total
68	Dee		20·5	20·4	865	22·1	59·3
		Average,	20·5	20·4	865	22·1	59·3

This somewhat too limited series of observations indicates—

1. That the male fish coming to the rivers throughout the season have a musculature somewhat poorer in solids than the female fish.
2. That in the upper reaches the percentage of solids is about the same as in the female fish.
3. That the nature of the change in the percentage of solids is the same as in the female fish.
4. That the testes are considerably poorer in solids than the ovaries.
5. That there is in the testes a more marked increase in the percentage of solids than in the ovaries in October and November.
6. That in the estuary fish, as regards the amount of solids in the muscle per fish of standard length, there is a marked increase in July and August, and a decrease in October and November.
7. That throughout the season the amount of solids in the muscle is smaller in the fish in the upper reaches than in those at the mouths of the rivers, and that the difference becomes more and more marked as the season advances.

TABLE V.

Solids of Muscle.

	Estuary.	Upper Water.
May and June.	1809	1774
July and August.	2452	1592
October and November.	1470	865

8. That the amount of solids in the testes per fish of standard length steadily increases in fish in the estuaries and in fish in the upper waters throughout the season, and that the increase is even greater in the sea than in the upper reaches. But since only one upper-reach fish was examined in October and November, too much stress cannot be placed upon this.

TABLE VI.

Solids of Testes.

	Estuary.	Upper Water.
May and June.	2·72	8·00
July and August,	4·73	18·60
October and November,	66·00	59·30

A balance of the loss of muscle solids and the gain in the solids of the testes may be struck as follows :—

A. Balance to August.

		Grms.
Solids in muscle of estuary fish from May to August,	-	2130
Solids in muscle of upper-water fish in July and August,	-	1592
Loss,	-	538
Solids in testes of estuary fish from May to August,	-	3·7
Solids in testes of upper-water fish in July and August.	-	18·6
Gain,	-	14·9

If 14·9 grms. of the solids of the flesh go to the testes, 523 grms. will be available as a source of muscular energy.

B. Balance to November.

		Grms.
Solids in muscle of estuary fish from May to August,	-	2130
Solids in muscle of upper-water fish in Oct. and Nov.,	-	865
Loss.	-	1265
Solids in testes of estuary fish from May to August,	-	3·7
Solids in testes of upper-water fish in Oct. and Nov.,	-	59·3
Gain,	-	55·6

If 55·6 grms. of the solids of the muscle go to the testis, then 1209 grms. are available for muscular energy.

The supply of solids, over that required for the construction of the testes, which is thus available for muscular energy, is considerably greater in the case of the male than of the female fish.

KELTS.

Table VII. gives the amount of solids per cent. and in fish of standard length in muscle and ovaries in four of the kelts received in the spring of 1897 :

TABLE VII.
Solids in Kelts.

No.	Muscle. Per Cent.		Muscle. Total per Standard Fish.	Ovaries. Per Cent.	Ovaries. Total per Standard Fish.
	Thick.	Thin.			
80	21·87	22·30	1021	14·14	6·20
81	23·79	25·85	1178	14·68	9·41
82	20·96	20·78	944	17·13	14·19
83	18·62	17·69	640	12·24	7·32
Average,	21·28	21·65	946	14·55	9·28

The length of these kelts (see **Table** p. 74) indicates very clearly that they belong to the large late-coming fish. They cannot, therefore, be compared with the fish in the upper waters in October and November, but should rather be compared with the estuary fish of these months.

This table shows that :

1. The percentage of solids in the muscle is slightly less than in the unspawned fish in the upper reaches in October and November, markedly less than in the unspawned fish at the estuaries in these months.

2. The amount of solids per fish of standard length is no less, perhaps rather greater, than in the unspawned fish in the upper reaches in October and November, but markedly less than the unspawned fish in the estuaries during these months. Hence the apparent increased size of the so-called well-mended kelt is in part, at least, due to increase in the water, and not in the solids of the muscle.

3. The percentage amount of solids in the ovaries is very much smaller than in fish ascending the rivers to spawn.

4. The solids of the ovaries per fish of standard length are markedly smaller in amount than in fish ascending the river to spawn.

8.—CHANGES IN THE FATS OF MUSCLE, GENITALIA. AND OTHER ORGANS OF THE SALMON IN FRESH WATER.

By D. NOËL PATON, M.D.. F.R.C.P.Ed.

Throughout the animal kingdom the material required for the evolution of energy is to a large extent stored in the body as fats. As is well known, the combustion of one gramme of fats liberates about twice as much energy as the combustion of a corresponding amount of proteids or carbohydrates. In the salmon the ova are largely composed of an oily fluid very rich in fats, and hence for the growth of the ovaries a large store of fats in the body is necessary.

In this section we have to consider where and how this fat is stored and to what extent it is used as a source of energy and to what extent for the growth of the genitalia.

METHODS.

A portion of the organ—usually about 30 grms.—was preserved in spirit until it was analysed. The spirit was then poured into an evaporating basin and the organ finely powdered in a mortar, added to the spirit, and slowly dried over a water bath. When it had not powdered easily at first, it was again powdered after drying. The muscle readily broke up into very fine small fibrils, while the ovaries and livers were readily reduced to a uniform powder. The powder was placed in a filter paper, the mortar and basin being carefully washed with ether and the ether added to the powder, and was then extracted in a Soxhlet apparatus for two days. If by the end of this time the ether was not absolutely colourless, the process was continued until it became colourless. The ether was distilled off and the fat dried at 100° C. and weighed. By the preliminary heating with alcohol the extraction of lecithin was facilitated.

This method was used in preference to the more recent procedure of Dormeyer (Pflüger's Arch. Bd. 65, p. 90), because the residue after extraction of fats was required for the estimation of proteids, phosphorus, and iron. To have put aside separate specimens for the analysis of each of these would have increased the accumulation of material to an unmanageable extent. As it was, no less than six or seven bottles of material were put aside for each fish.

In face of Dormeyer's severe strictures upon Soxhlet's method it was necessary to test it carefully against the method devised by the former.

Dormeyer found that with horse flesh Soxhlet's method of extraction failed to remove a very considerable portion of the fats, and he recommends that after an initial extraction by Soxhlet's method the residue should be submitted to artificial gastric digestion, by which the muscle fibres are dissolved and the fats liberated. These may then be removed by shaking the fluid with ether.

The muscle, ovaries, and testes of the salmon after preservation in spirit are readily reduced to a very fine state of subdivision, and have not the tough, fibrous consistence of horse muscle. The ether in Soxhlet's apparatus has thus a much better chance of completely recovering the fats. But to test the method against Dormeyer's, the following experiment was performed on Salmon 44 :—

Ovaries.—38 grms., treated in the usual way yielded 3·841 grms. of extract in Soxhlet's apparatus after two days' extraction. The residue was subjected to peptic digestion for 12 hours, and yielded a brown fluid and a copious residue. The fluid was filtered off, and the residue and paper well washed with ether. The filtrate was extracted with ether in a separation funnel. On distilling and evaporating the ether, 0·022 grm. of residue was obtained—·54 per cent. of the total extract.

Muscle.—Of the thick 44 grms. and of the thin 35 grms. were treated in the same way. The residue after digestion was very small in amount. The following are the results :—

			Extraction by Soxhlet's Method.	Subsequent Extraction by Dormeyer's Method.
Thick,	-	-	3·477	0·008
Thin,	-	-	5·158	0·007

In thick, 0·23 per cent. of ether extract.
In thin, 0·13 per cent. of ether extract.

Soxhlet's method as employed by us may thus be considered to give quite satisfactory results in the case of the muscle, and somewhat too low results in the case of the ovaries, but the difference is so small that in calculating the percentage of fats in the ovaries the Soxhlet's extraction gives less than 0·1 per cent. less than Dormeyer's method. Soxhlet's method, 10·107 per cent.; Dormeyer's method, 10·166 per cent. For our purpose the accuracy of the method is amply sufficient.

The ether extracts after weighing were preserved in small flasks, and in some the amount of fatty acids were determined by Kossel and Oebermüller's method (Ztsch. f. phys. Chem. Bd. xiv. 599), while in others the lecithin was estimated in the usual way.

AMOUNT OF FATTY ACIDS.

The following determinations show the amount of fatty acids present in the ether extract of the muscle, and indicate that it is very largely composed of ordinary fats :—

TABLE 1.

No.	Per Cent. of Fatty Acids in Ether Extract.	
	Thick.	Thin.
32	86·5	86·0
66	89·5	88·5
74	91·7	92·2

FEMALE SALMON, 1896.

Table II. gives the amount of fats in per cent. and in fish of standard
length in the "thick" and "thin" of the muscle, in the whole muscle,
and in the ovaries.

TABLE II.

ESTUARY FISH.

May and June.

No.	River.	Muscle.			Ovaries.	
		Per cent. Thick.	Per cent. Thin.	Total per Standard Fish.	Per cent.	Total per Standard Fish.
A 16	Dee	10·70	16·9	732	9·04	8
20	Helmsdale . .	10·90	17·1	838	9·40	13
25	Helmsdale . .	8·14	14·6	641	9·45	16
27	Dee . . .	11·20	21·3	861	8·66	12
	Average,	10·23	17·9	768	9·14	12
B 15	Spey . . .	3·96	7·27	287	11·60	12
17	Dee	8·85	12·00	329	8·50	7
29	Dee	7·10	15·20	510	7·90	11
	Average,	6·64	11·5	375	9·30	10
	Average of A & B,	8·69	14·9	599	9·17	11

July and August.

No.	River.	Per cent. Thick.	Per cent. Thin.	Total per Standard Fish.	Per cent.	Total per Standard Fish.
36	Dee	8·5	17·0	688	6·9	18
40	Dee	10·6	17·8	867	6·9	7
45	Spey . . .	12·6	17·3	862	8·7	13
51	Spey . . .	10·2	14·2	782	9·9	19
55	Dee	7·5	15·3	653	12·0	35
	Average,	9·82	16·8	770	8·9	18

October and November.

No.	River.	Per cent. Thick.	Per cent. Thin.	Total per Standard Fish.	Per cent.	Total per Standard Fish.
65	Spey . . .	7·03	9·8	562	9·73	151
73	Dee	5·07	12·1	403	10·00	159
74	Dee	4·15	8·4	313	9·64	126
	Average,	5·41	10·2	426	9·79	145
72	Dee . . .	15·1	20·0	1109	8·05	7·9
79	Helmsdale . .	12·6	20·2	962	10·80	6·8
	Average,	13·3	20·1	1035	9·42	7·3

TABLE II.--*Continued.*

UPPER WATERS.

May and June.

No.	River.	Muscle.			Ovaries.	
		Per cent. Thick.	Per cent. Thin.	Total per Standard Fish.	Per cent.	Total per Standard Fish.
12	Spey	6·87	9·4	418	12·5	16
21	Helmsdale	8·90	13·4	656	12·2	26
31	Spey	6·02	10·2	384	10·8	32
32	Dee .	5·62	8·1	383	10·5	43
11	Spey	6·63	10·5	432	6·8	11
	Average,	6·81	10·3	448	10·5	29

July and August.

37	Spey	9·31	15·8	581	11·7	41
42	Spey	5·75	10·0	394	9·9	56
43	Helmsdale	6·35	10·1	388	10·6	37
49	Helmsdale	7·03	13·2	540	11·2	52
	Average,	7·11	12·2	476	10·8	46

October and November.

62	Helmsdale	2·50	3·13	107	9·55	193
63	Helmsdale	3·34	4·21	131	9·62	177
64	Dee .	2·68	6·77	167	10·00	211
66	Helmsdale	6·66	13·60	335	9·83	213
67	Helmsdale	2·37	5·59	121	9·76	197
69	Helmsdale	3·72	6·90	175	7·56	204
70	Dee .	0·62	4·16	59	6·90	141
	Average,	3·11	6·34	159	9·03	204

From this Table the *average* percentage of fats may be calculated :

TABLE III.

	Estuary.			Upper.		
	Thick.	Thin.	Ovaries.	Thick.	Thin.	Ovaries.
May and June .	10·2	17·9	9·1	6·8	10·3	10·5
July and August .	9·8	16·8	8·9	7·1	12·2	10·8
Oct. and Nov. .	5·4	10·2	9·8	3·1	6·3	*9·0

The somewhat smaller percentage may be due to the more difficult extraction of fats from the ripe ova, which, after preservation in alcohol, become hard and difficult to reduce to a powder.

The *average* amount of fats, per fish of standard length, is given in Table IV.:—

TABLE IV.

	Ovaries.		Muscle.	
	Estuary.	Upper.	Estuary.	Upper.
May and June	12	29	768	448
July and August	18	46	770	476
October and November	145	204	426	159

These results closely correspond to the results obtained in the study of the changes in the total solids, but the proportionate changes are even greater.

1. In the fish coming to the estuaries the percentage of fat and the fat per fish of standard length in the muscle is practically the same from May on to August. The fish coming to the rivers in October and November have a markedly smaller percentage of fat, and a markedly smaller amount per standard fish— about three-quarters of the original amount.

2. In the ovaries, in fish coming from the sea, the percentage of fat does not alter much throughout the season. In October and November it is slightly higher than in the earlier months. The amount of fat in the ovaries per fish of standard length undergoes no very marked changes from May to August—an increase of from 12 to 18 grms. But in October and November there is a rise to 145 grms. This is an eight-fold increase.

3. In the upper reaches, the fish in July and August show a slight increase in the percentage of fat and in the fat per fish of standard length in the muscle, when compared with the May and June fish. This is probably to be explained by the constant arrival of fresh fish from the sea. In October and November the percentage of fat falls to less than a half of the amount in the earlier months, while the fat per fish of standard length shows a fall of about one-third.

4. As compared with the fish in the estuaries, the percentage of fat, and the fat per fish of standard length, in these upper-water fish is very much smaller in amount.

It should perhaps be noted that there is some indication of a division of the upper-water fish, especially from May to August, into two classes, one richer and the other poorer in fat. Thus in May and June number 21 has a much greater amount of muscle fat than the other fish of the class.

In July and August 37 and 49 have an average of 560 grms., while 42 and 43 have only 391 grms.

At first sight this would seem to indicate that some of these fish have come up earlier than others, and have lost a greater quantity of fat from their muscles. But a comparison of these figures with the total solids and with the proteids opposes the idea that such a distinction can be made.

5. In the ovaries the percentage of fat is slightly higher in upper-water fish than in the fish from the estuaries, except in October and November. As already pointed out, this is probably an apparent, and not a real difference.

6. Throughout the whole season the fat per fish of standard

G

length in the ovaries of upper-water fish is about double that in estuary fish. The amount shows a small, though marked, increase **from** May **to August, and an increase** of four and a **half times in October and November as compared with** the earlier months.

A balance between fat lost from the **muscle and fat gained by the** ovaries shows the following results:—

(*a*) *Balance to August.*

Muscle.

			Amount of Fat in grms.
Average in Estuary,	-	May to August, -	- 770
„ in Upper Waters,	-	July and August, -	- 478
		Loss,	- 292

Ovaries.

Average in Estuary,	-	May to August,	- 15
„ in Upper Waters,	-	July and August, -	- 46
		Gain,	- 31

Of muscle fat there goes to ovaries,   -   - 31 grm.
„   „   „ there is available for energy,   - 261 grm.

(*b*) *Balance to November.*

Muscle.

Average in Upper Waters,	-	October and November,	159
„ in Estuary,	-	May to August,	- - 770
		Gain,	- 611

Ovaries.

Average in Upper Waters,	-	October and November,	204
„ in Estuary,	-	May to August,	- 15
		Loss,	- 189

Of muscle fat there goes to ovaries,   - 189 grm.
„   „   „ there is available for energy,   - 422 grm.

It is thus manifest that the fat which the salmon has stored in its muscles when it leaves the sea is not only amply sufficient to yield all the fat required for the fats of the growing ovary, but also abundantly sufficient to yield energy for an enormous amount of muscular work. When **the changes** in the proteids **of** the muscle have been considered, the **part played** by each **of these** in the evolution of energy will be discussed.

INTERNAL FAT.

(*a*) *Fat of Pyloric Appendages.*

Not only has the salmon leaving the sea this **store of** fat in its muscles but **it** has also fat stored **in** the abdominal cavity round the intestine and **in** the liver.

In the salmon the visceral **fat is** chiefly collected **round the pyloric** appendages According to Miescher Ruesch (*loc. cit.*), **p. 179, by the** beginning of August the intestinal fat has almost disappeared. **This observation,** of course, applies only to fish in the upper waters. The **state of the** intestinal fat in fish in the lower waters has not, so far as I am aware, been studied.

In every fish after Number 20 a note of the fat on the appendages was kept, and the following table gives the results of these observations:—

TABLE V.

	Estuary.				Upper Water.			
	No. of Fish.	Much Fat.	Little Fat.	No Fat.	No. of Fish.	Much Fat.	Little Fat.	No Fat.
May and June, ..	8	6	1	1	4	1	1	2
July and August, ..	13	9	4	0	5	2	3	0
October and November, ..	7	*3	3	1	8	0	0	8

* Two of these, 72 and 79, were "winter salmon" and are not included in the percentage table below.

This gives the percentage of fish as follows:—

TABLE VI.

	Estuary.			Upper Water.		
	Much Fat.	Little Fat.	No Fat.	Much Fat.	Little Fat.	No Fat.
May and June, ...	75	12	12	25	25	0
July and August, ...	39	51	0	40	60	0
October and November, ..	20	60	20	0	0	100

In order more accurately to show the extent of the change in the intestinal fat a series of analyses were made from fish at the mouth and from fish in the upper reaches of the rivers throughout the season. The examples selected for this were average specimens.

METHOD.

The pyloric appendages with the upper part of the intestine **were** preserved in spirit. They were then chopped up, and repeatedly extracted with hot alcohol, the alcohol being filtered hot. The residue was then repeatedly extracted with ether. The alcoholic residue **was** dried over the water bath, and repeatedly extracted with ether, the ether being united with that **from** the residue. The ether was distilled off, **and** the fats weighed in **the** usual manner.

Table VII. gives the results of these analyses :—

[TABLE.

TABLE VII.

FATS OF PYLORIC APPENDAGES.

	Estuary Fish.			Upper-Water Fish.		
	No.	Total Fat.	Fat per Standard Fish.	No.	Total Fat.	Fat per Standard Fish.
	27	11·741	28·4	31	2·446	5·73
May to August, ……	45	14·085	33·4	32	6·930	11·z
	41	20·850	58·9	43	4·987	17·3
	73	2·108	2·88	69	0·506	1·21
	74	3·812	5·06	70	0·435	1·50
Oct. and Nov…	72	47·420	67·2			
	79	Enormous quantity of fat, more than in 72. This was lost in analysis by breaking of flask.				

These figures indicate:—

1st, That in the early part of the season there is a marked difference as regards the amount of intestinal fat in fish at the mouth and in fish in the upper reaches.

2nd, That in October and November the amount of fat on the appendages of the fish coming to the mouth of the river has greatly diminished.

3rd, That the "winter salmon" appearing at this time, 72 and 79, *i.e.*, fish not ready to spawn till the next season, have an enormous accumulation of fat on the intestine.

Taking the averages from May to August of estuary fish and fish in the upper waters, it is found that 28·5 grm. of fat per standard fish is used up.

From these results **it** *would appear that this intestinal fat is the first to be drawn upon,* **and that it is** *used up more rapidly than the muscle fat.*

(b) *Liver Fat.*

In many fish, such as the cod, the great accumulation of fat occurs in the liver. I have shown elsewhere ("Journal of Physiol.," vol. xix, p. 172) that as much as 67 per cent. of fat is found in the liver of the Gadidae.

In the female salmon there is never such an accumulation of liver fat. In 1895 the amount of fat in the liver of several fish was determined, and no marked change was observed from July to October in fish in the estuary.

In 1896 the liver fats were determined in four average fish :—

TABLE VIII.

	Estuary Fish.			Upper-Water Fish.	
No.	Per Cent.	Fat per Standard Fish.	No.	Per Cent.	Fat per Standard Fish.
27	17·90	29·7	69	3·30	4·66
45	3·65	29·5	70	3·53	3·94

The livers of the two exceptional fish 72 and 79 from the mouth of the river in November yielded the following amounts of fat :—

72	Per Cent, 12·56.	Per Standard Fish, 21·1.
79	„ „ 5·76.	„ „ „ 10·2.

These figures show that the fats stored in the liver while the fish is feeding in the sea are to a great extent lost during the sojourn in fresh water. As much as 20 grm. of fat per fish of standard length may be given off from the liver.

FEMALE SALMON, 1895.

Table IX. gives the analyses of fats in the North Esk salmon of 1895. It agrees with Table II. except as regards the ovaries. A greater percentage and actual amount of fat is present in these North Esk fish in July than in the 1896 fish of the same month :—

TABLE IX.

No.	Date.	Muscle.			Ovary.	
		Per Cent.		Total per Standard Fish.	Per Cent.	Total per Standard Fish.
		Thick.	Thin.			
10	23.7	9·4	14·3	625	10·3	51
12	,,	12·7	14·9	1015	10·7	47
	Average,	11·0	14·6	820	10·3	49
29	26.10	6·4	10·3	449	8·2	59
31	28.10	9·6	18·7	795	9·5	114
39	12.12	5·1	1·9	212	10*	210
	Average,	7·0	13·3	465	9·2	127

* Analysis lost. Average from others.

MALE FISH.

In the male salmon the testes are comparatively poor in fats. For their development there is no need of the same storage of fats as in the case of the growth of the ovaries. It is thus a matter of very considerable interest to ascertain whether in the male fish the same accumulation of fat in the muscle occurs as in the female, and whether this fat is used up to the same extent.

Table X. gives the results of the analyses of the fat in male fish during 1896 :—

[TABLE.

TABLE X.

ESTUARIES.

May and June.

No.	River.	Muscle.			Testes.	
		Per Cent.		Total per Standard Fish.	Per Cent	Total per Standard Fish.
		Thick.	Thin.			
13		5·07	13·1	452	3·2	·34
19		11·5	17·7	730	5·6	·87
	Average,	8·73	15·4	591	4·4	·6

July and August.

56		10·7	17·7	865	3·7	·73
59		12·3	19·1	980	2·9	1·04
	Average,	11·5	18·4	922	3·3	·88

October and November.

71		4·0	9·6	326	3·3	7·16
75		2·8	5·6	183	2·9	9·00
	Average,	3·4	7·6	254	3·1	8·08

UPPER WATER.

May and June.

26		11·1	15·7	762	4·8	2·02
34		6·1	9·6	359	3·3	1·45
	Average,	8·6	12·6	561	4·0	1·73

July and August.

38		8·2	10·7	524	3·3	1·46
39		8·4	14·4	523	1·9	1·87
54		4·5	6·9	238	1·9	3·91
	Average,	7·0	10·6	428	2·3	2·41

October and November.

| 68 | | 2·2 | 3·4 | 103 | 2·3 | 6·34 |

The average results of this table may be given as follows:—

A. *Muscle.*

TABLE XI.

PER CENT. OF FATS.

	Estuary.		Upper Water.	
	Thick.	Thin.	Thick.	Thin.
May and **June**,	8·7	15·4	8·6	12·6
July and August,	11·5	18·4	7·0	16·0
October and **November**,	3·4	7·6	2·2	3·4

TABLE XII.

FAT PER FISH OF STANDARD LENGTH.

	Estuary.	Upper Water.
May and **June**,	591	561
July and August,	922	428
October and November,	254	103

B. *Testes.*

TABLE XIII.

FAT IN PERCENTAGE AND PER FISH OF STANDARD LENGTH.

	Estuary.		Upper Water.	
	Per Cent.	Per Standard Fish	Per Cent.	Per Standard Fish
May and **June**,	4·4	0·6	4·0	1·73
July and August,	3·3	0·88	2·3	2·41
October and **November**,	3·1	8·08	2·3	6·34

The fat balance in male fish is given below :—

A. *Balance to August.*

Muscle.

	Grms.
Average in estuary fish from May to August,	756
Average in upper-water fish in July and August,	428
Loss,	328

Testes.

	Grms.
Average in estuary fish from May to August,	0·74
Average in upper-water fish in July and August,	2·41
Gain,	1·67

The testes thus accumulate 1·67 grms. of fat, and thus 326 grms. of fat are available as a source of energy.

B. *Balance to November.*

Muscle.

	Grms.
Average in estuary fish in May and August,	756
Average in upper-water fish in October and November,	103
Loss,	819

Testes.

	Grms.
Average in estuary fish in May and August,	0·74
Average in upper-water fish in October and November,	6·34
Gain,	5·46

The testes take up 5·46 grms. of fat, and thus 813·5 grms. are left as a source of energy.

On comparing these figures with the results obtained from female salmon it will be seen that *the accumulation of fat in the muscles is as great in the male as in the female, and that the fat is used up to quite as great an extent. On the other hand, the accumulation of fat in the testes is trifling in amount, and thus the conclusion is indicated that in the male the utilisation of fat as a source of energy is much greater than in the female. This is especially marked in the later months.*

INTERNAL FAT.

A. *Pyloric Appendages.*

The storage of fat on the pyloric appendages of male fish was not investigated by analyses.

Some observations were, however, made upon the fats of the liver during the season 1895.

Table XIV. gives the results of these :—

[TABLE.

TABLE XIV.

No.	Date.	Solids.	Fats.
4	6.7	39·5	12·9
5	10.7	35·9	19·7
6	10.7	38·1	24.5
19	8.8	42·6	28·0
40	12.12	20·3	1·5
43	24.12	• 18·9	3·1

All the fish except 40 and 43 were from the estuaries, but these two fish were taken from the spawning beds.

KELTS.

Table XV. gives the amount of fats in the muscles and ovaries of the female kelts analysed :—

TABLE XV.

No.	Muscle.			Ovaries.	
	Per Cent.		Total per Standard Fish.	Per Cent.	Total per Standard Fish.
	Thick.	Thin.			
80	2·27	3·95	125	2·14	·94
81	1·79	8·05	278	2·28	1·46
82	1·86	2·98	96	1·33	3·67
83	·92	1·09	31	1·64	·98
Average,	2·13	1·02	133	2·02	1·76

These observations bring out the following points :—

1. The per cent. of fats in the muscle is only slightly lower than in the upper-water fish of October and November, but markedly lower than in the estuary fish of these months.*

2. The same applies to the amount of fats per fish of standard length.

3 The percentage of fats in the ovaries is markedly lower than in any of the fish ascending the rivers.

4. The amount of fats per standard fish is still lower. -

---

* It has been shown (p. 92) that it is with these estuary fish that the kelts must be compared.

9.—MICROSCOPICAL OBSERVATIONS ON MUSCLE FAT IN THE SALMON.

By S. C. MAHALANOBIS, B.Sc., F.R.M.S.

The chemical observations on the changes in the fats of the salmon during its sojourn in fresh water have shown that the fish leaves its marine feeding ground with the muscles loaded with fat, and that this fat gradually diminishes in amount, being in part transmitted to the ovary and in part used up as a source of energy.

The object of the present enquiry is to investigate more fully the nature of this change.

On this subject Miescher Ruesch says (p. 186) :—*

"That the lateral trunk muscles are the actual source of material both for the nourishment of the animal and for the ripening of the genitalia, is rendered evident by the microscope. Even the winter and spring salmon show a sometimes more, sometimes less, well marked series of fat droplets chiefly between the fine, cross-striped, elementary fibrillæ of the unequally thick muscle fibres, especially in the thinner ones, such as we recognise as a sign of the so-called degeneration of muscle fibres. The amount of these fat droplets increases to mid-summer, when the ovary begins to grow more rapidly, and may lead to many fibres becoming opaque. A separate thin muscle plate which lies along the side of the body just under the skin degenerates most markedly. On the other hand, all the remaining muscles of the breast, belly, back and anal fins, of the jaw and hyoid bone, the upper and lower longitudinal muscles (Längsmuskel), and the tail muscle, in the stricter sense, continue, so to speak, fully intact and free of fat."

Again on p. 208 he says :—" . . . the trunk muscles, which already in the salmon in March, nay even in the winter salmon (December), show clear traces of commencing fatty degeneration (Fettentartung)."

On p. 215, in describing the condition of the fish on the spawning beds, he says :—The flesh of the trunk is entirely opaque, whitish, and entirely filled with fat droplets." As will be seen presently, our observations do not agree with this last description.

Miescher then goes on to develop a theory of the liquidation and fatty degeneration of the muscles. After referring to the influence of diminished vascular supply and diminished respiration upon the metabolism, he concludes that the changes in the muscle are caused by the diminution in the blood supply.

It is thus a matter of very considerable importance to re-investigate the microscopic appearances of the muscle fibres and to determine how far Miescher is correct in his conclusion that a fatty degeneration occurs.

It has already been shown (p. 93) that the salmon feeds during the time it remains in the sea, and that a store of fat appears in the muscles, which is evidently used up for the nutrition of the animal, as well as the ripening of the generative organs, during its sojourn in the river.

The results of the chemical examination clearly point to two facts :— First, both at the mouth of the river and in upper water there is a fall in the amount of muscle fat from May to November, the amount of fat

* Statistische und biologische Beiträge, "Zur Kenntniss von Leben des Rheinlachses im Susswasser."

in November fish being about half that in May fish. Second, there is an enormous difference between the average amount of fat in fish at the mouth and the average in those in the upper water during the later months of the year.

The exact nature of this change in the amount of fat can only be satisfactorily understood when the results of chemical examination are supported by histological evidences.

I.—METHODS.

Pieces of the lateral trunk muscle both thick and thin, *i.e.* dorsal and abdominal, from each fish, were fixed in (*a*) sat. sol. of corrosive sublimate, and (*b*) in 1 per cent. solution of osmic acid for 24 hours, then thoroughly washed, hardened in alcohol in the usual way, and quickly dehydrated. They were embedded in paraffin, and longitudinal as well as transverse sections of 5 to 8 microns thickness were cut. Various solvents of paraffin, *e.g.* cedar oil, xylol, clove oil, benzole, were tried before embedding, and xylol was found to answer best in preserving the fat. The sections were mounted on albuminised slides to afford every facility for the preservation of fat. It was found by experiments that fixing tissues in reagents other than osmic acid, and afterwards treating the sections with osmic acid, did not give satisfactory results. Sections fixed in corrosive sublimate were stained with some suitable double stain such as hæmatoxylin and eosin or methyl-blue and eosin. The fat in these sections was, to a great extent, washed out, and they then served for comparison with the osmic acid sections, and helped in the study of the general change in the muscles.

The reduction of osmium in sections fixed in osmic acid revealed the presence of fat globules in a most perfect manner, and it only needed some single stain, *e.g.* eosin, to bring out other details. Tissues fixed in osmic acid are rather difficult to stain, but eosin seemed to answer the purpose well.

II.—RESULTS OF EXAMINATION OF THE SECTIONS.

A.—THOSE FIXED IN PERCHLORIDE OF MERCURY.

1. *Muscle Fibres.*—The size of the muscle fibres varied very much, from 50 to 167 microns in diameter, the largest size being in early fish from the mouth of the river. In two specimens caught at the mouth in August, the fibres were about 134 microns in diameter. In those from the upper reaches the size of the fibres varied from 50 to 100 microns in diameter, the smallest size being in the thin or abdominal muscle of specimen No. 63, a fish caught in October.

2. *Striation.*—In longitudinal sections the transverse striations were well seen in fibres in the centre of a bundle, whereas the fibres at the periphery showed, in many cases, a tendency to longitudinal cleavage, and there the transverse striation was obscure. This longitudinal cleavage was most pronounced in fish from the mouth of the river.

3. *Muscle Fibrilla.*—The size of the individual fibrils in muscle fibres seemed fairly constant in all specimens, being of 1·5 microns to 1·8 microns in diameter. The number of Conheim's areas varied with the size of the fibres.

4. *Nucleus.*—The longitudinal sections showed well-marked, rather elongated nuclei. No change was noticed in the different specimens except that in No. 79 they were very prominent, but there was no indication of any active proliferation having taken place.

5. *Amount of Fat.*—The amount of fat in specimens fixed in corrosive sublimate could only be approximately ascertained by the honeycomb-like appearance of the empty spaces in the connective tissue left by the fat cells, and by a comparison with the osmic acid preparations. In the *perimysium* this layer of empty spaces shows an average of ¼mm. in

thickness in the case of fish leaving the sea early in the year; whereas towards the end of summer and in autumn I find an average of $\frac{1}{2}$ mm. thick. For the amount of fat in the *endomysium*, as well as the intracellular fat, we must depend on osmic acid preparations.

B.—THOSE FIXED WITH OSMIC ACID.

1. *Fat between Muscle Fibres.*—In fish coming from the sea there is an abundance of fat cells between the muscle fibres, but in fish that have been for some time in the river this fat has almost entirely disappeared. Fig. 4 is from an upper-water fish, hence even in this longitudinal section we notice a marked absence of *intercellular* fat, whereas Fig. 5, being from No. 79 (a fish fresh from the sea), shows, even in the transverse section, presence of abundance of fat.

2. *Fat Inside the Muscle Fibres.*—Treatment with osmic acid explained the cause of longitudinal cleavages in the muscle fibres, and revealed the presence of fine granules of fat along the cleavage lines between bundles of fibrils, also between individual fibrils (Fig. 1). In transverse sections also these granules can be seen lodged between the fibrils. (Fig 2.)

To make sure that these *interfibrillary* granules of fat were not just small particles broken up from connective tissue-fat and scattered over the sections during washing, etc., every slide, during the whole process, was kept in a vertical position with a marked end always upward; so that all flow of particles, if there were any, would be in one direction only; but the specimens show a very uniform distribution of these granules.

The amount of this *intracellular* fat varies very much at different periods. In early fish the amount is much greater and the granules much larger than in late fish.

In the fish leaving the sea this accumulation of fat in the fibres sometimes reaches an enormous amount, and a thick layer occurs under the sarcolemma. This will be evident from a comparison between Fig. 2 and Fig. 3, the former being from a late fish, 69, and the latter from No. 79.

Chemical analyses showed the amount of fat in these fish to be, in per cent :—

	Thick.	Thin.
69	3·7	6·6
79	12·6	20·2

Summing up the results of the examination of the specimens fixed in corrosive sublimate and of the corresponding ones in osmic acid, we find that the evidence of the microscope tallies with the result of the chemical examination, and points not only to a change in the amount of muscle fat, but also, to some extent, in the nature of its distribution, at different periods. The early fish at the mouth of the river have a much greater amount of fat in the muscles (both intercellular and intracellular) than the late fish in the upper reaches.

This diminution of muscle fat in the late fish may be due to a want of fresh accumulation or to an increasingly active removal. But the enormous difference in amount between a fish like No. 72 or No. 79 and one that is about to spawn, cannot be accounted for by either of these causes singly. There is, no doubt, more active removal of fat— probably slightly due to increased amount of work in going up the river—but mainly due to an export to the generative organs. A fish like No. 79 has a very small ovary and large amount of muscle, whereas a fish about to spawn has a very large ovary and small amount of muscle. This obviously points to the fact that in the late fish the ovary grows at the expense of the muscles. At the same time, it is evident that if during the growth and development of the sexual organs there had been fresh accumulation of fat—or, in other words, if the animal had been actively feeding—the equilibrium would have been maintained and the muscles would not have lost, at all events, to such an extent.

It would appear, then, that the fats taken in its food by the salmon in the sea accumulate between the muscle fibres and also inside the fibres between the fibrils, and that during the sojourn of the fish in the river these fats steadily diminish, being either used up as a source of energy by the muscle, or transported from the muscle to the growing ovaries. They do not bear out Miescher Ruesch's description of the condition of the muscles in the spawning fish, and they entirely oppose his view that anything of the nature of a fatty degeneration occurs.

The results of this investigation throw some light on the so-called "fatty degeneration of muscle." Endless controversy has raged to decide the question of direct formation of fat from proteids in the cell. The results of Pettenkofer and Voit's classical experiments, the evidence of the formation of fat from blood by maggots, and the fatty change in the ripening of cheese have been recently dealt with by Pflüger (*) who shows that the evidence is far from conclusive. The argument of "fatty degeneration of muscle" is the sheet anchor to the supporters of the theory of the proteid-origin of fat. But even with that their position is not secure, as has been pointed out by Dr Noël Paton (†). Pathologists usually fall into the error of depending on the evidence of the microscope only, without making thorough chemical investigation. But it has been pointed out by Krehl (‡) that, hearts showing the characteristic microscopic symptoms of fatty degeneration, may have less than normal amount of fat. In this investigation I had the advantage of comparing microscopical observations with the results of careful chemical examination made by Dr. Noël Paton.

A glance at the Figures 1, 2, and 3 will at once show that it would be rather rash to come to a conclusion depending on the evidence of the microscope alone. Here we have the appearance at least resembling the so-called "fatty degeneration" of muscle. But how can there be any fatty degeneration unless the fat is formed at the expense of the proteid molecules of the muscle fibres ? The simple fact that the extremely minute granules of fat are found to arrange themselve between the fibrils is no proof of their formation from the muscle substance. Fraser and Bruce (§) have described a somewhat similar appearance in the sections of the tibialis posticus muscle from a case of diabetic neuritis, and called it "disseminated interfibrillary fatty *degeneration*," suggesting, at the same time, the origin of the fat from the cement substance rather than the muscle fibrils. But non-utilization of fat by muscles, due to failure of trophic influence of nerves and want of functional activity, may lead to an accumulation of fat which might be mistaken for fatty degeneration. It has been shown here that microscopic appearances, such as are described by the pathologists as typical of fatty degeneration, may be found in conditions that are necessary in the economy of Nature. In our specimens, although careful chemical examination detects no diminution of proteid in the muscle substance—at all events nothing like the extent which would account for the enormous amount of fat—there is the pseudo-evidence of the microscope pointing to so-called "fatty degeneration."

It is interesting to note that in the African mud fish—Protopterus annectens—great accumulation of fat appears in the lateral

(*) Pflüger's Arch., 52, 1 and 239. 1892.
(†) Journal of Physiology. Vol. XIX. No. 3. 1896.
(‡) Deutsch. Arch. f. klin. Med., LI., 416. 1893.
(§) Edinburgh Medical Journal, October 1896.

muscles alongside the spinal axis in the tail, which serves as reserved material for the nutrition of the animal, as well as the formation of the generative products, while it passes into a torpid state during the dry season and encloses itself into a cocoon.

Parker (*) supposes, on the authority of Professor Ziegler, that in Protopterus the lateral muscles in the tail undergo fatty degeneration. He also notices a granular degeneration such as is described by Schneider (†) in the case of Petromyozon fluviatilis. The general appearance of the changes in the muscle fibre resembles that described by Fraser and Bruce (in the case already referred to), inasmuch as " the disintegration occurs in small islands in a muscular fibre." Parker thinks that in Protopterus there appears first a fibrillar change causing a loosening of the muscular substance, followed by a granular degeneration—which is probably the precursor of fatty degeneration.

But, as his observations were made only on specimens in the torpid state (and almost all obtained in July), it would be hardly satisfactory to come to any definite conclusion as regards the nature of the muscle-fat —without a comparison of the appearances of the muscle—before and after the torpid period, and also at different stages of the same.

In my specimens of the muscle of salmon (which, as already mentioned, were taken from fish of different times of the year) I did not notice any granular change.

It is probable that Protopterus passes into the torpid state with its muscles loaded with a store of fat, which is steadily consumed during its captivity. On this supposition the "fibrillar change" is quite explicable, as will be shown later on, in the case of the salmon.

On the assumption of actual degeneration of muscular tissue, Parker suggests the probability of the re-absorption of the degenerated products, in Protopterus, lamprey and salmon, being brought about by the agency of wandering leucocytes, after the manner described by Metschnikoff (‡). But, at the same time, he recognises the difficulty in the acceptance of such an explanation ; as, Loos (§) has pointed out that Metschnikoff's researches regarding the part played by leucocytes in the absorption of tadpole's tail are not conclusive. Whereas, in the case of the salmon, the explanation is rendered invalid by the fact of its depending on data that, I contend, still remain not proven.

" Fatty degeneration " is at present a misnomer. This investigation and similar other observations strongly suggest that, in many cases, the so-called fatty degeneration is a mere fatty infiltration due to increased accumulation of fat from diminished utilisation in the tissues. Bearing in mind the result of chemical investigation, the appearance in the Figures 1, 2, and 3, at all events, should be described as interfibrillary *infiltration* of fat. It has already been stated that Figure 3 is from a fish fresh from the sea—one that had been actively feeding, and consequently its blood and lymph were rich in fat, whence, in all probability, the muscle cells absorbed fat and stored it between the fibrils. We know that a dense network of capillaries surrounds the muscle fibres, and although no capillaries enter the fibres, there are lymph spaces surrounding Conheim's areas and communicating with those beneath the sarcolemma. As already pointed out, the fat granules in fish leaving the sea are more crowded immediately under the sarcolemma (Fig. 3.). Figure 2 is from a late fish—one that had been actively using up the reserve fat, hence the transverse section of

(*) "Anatomy and Physiology of Protopterus Annectens." Transactions of the Royal Irish Academy. Vol. XXX., Part III., p. 207
(†) Beiträge zur vergl. Anat. u. Entwickelungsgeschichte d. Wirbelthiere, Berlin, 1879.
(‡) "Untersuch ueb. die mesodermalen Phagocyten einiger Wirbelthiere."—Biol. Centralblatt Bd. III.
(§) Biol. Centralblatt, IX., 1889.

FIG. 1.

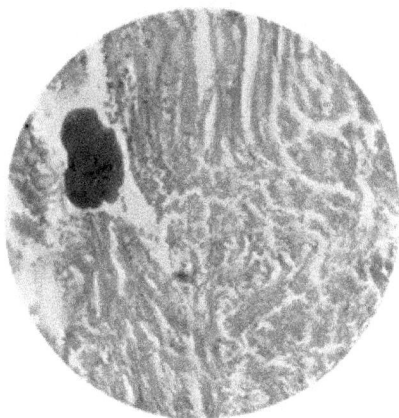

× 300.

PLATE II

FIG. 2.

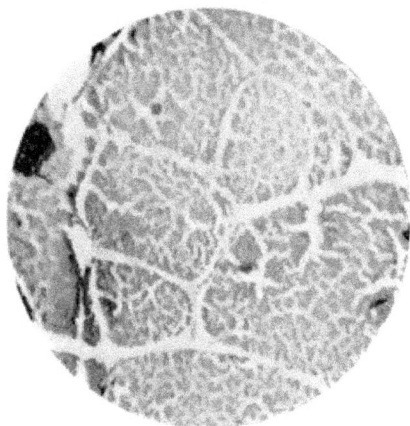

× 600.

FIG. 3.

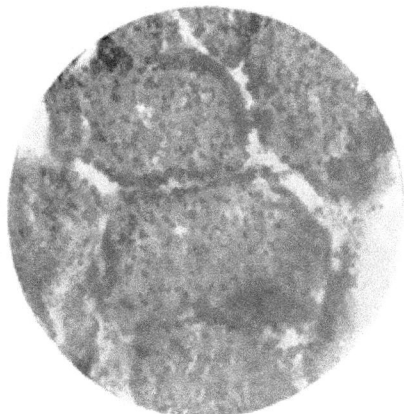

× 600.

PLATE III.

FIG. 4.

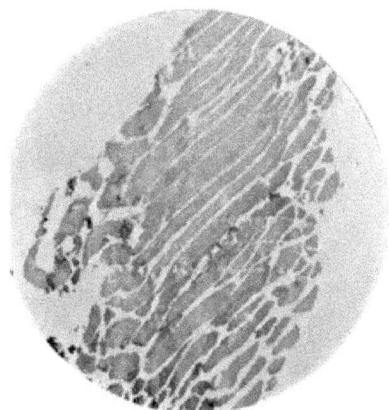

× 50.

FIG. 5.

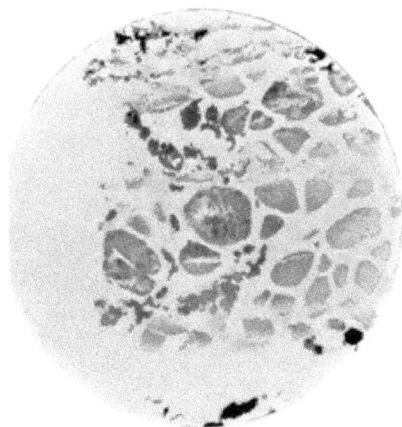

× 100.

PLATE IV.

FIG. 6.

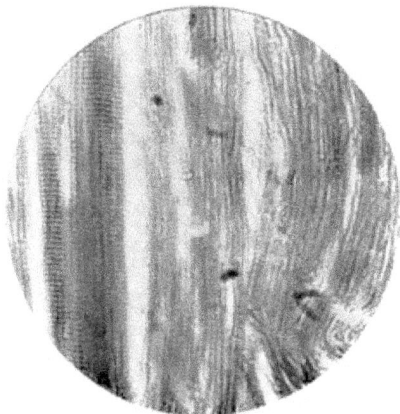

× 600.

FIG. 7.

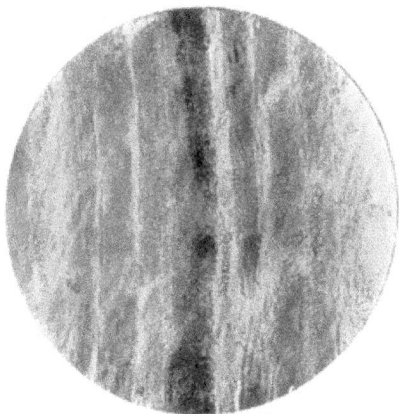

× 600.

the fibres does not look nearly so crowded with it; the fat granules are much smaller and more scattered, and the masses at the periphery of the fibres are all used up.

A comparison between the appearances of the Figures 2 and 3 would suggest the idea of a secreting cell during activity in the case of the former, and at rest in the latter. But it is a mere analogy and cannot be pushed far, as that would involve a formation of fat by the protoplasm of the cell.

As generally stated by pathologists, and also found in my specimens, the fibres that undergo such fatty changes show indications of disappearance of the cross stripes. This loss of transverse striation is usually concomitant with the appearance of marked longitudinal cleavages between bundles of fibrils and also individual fibrils. Hence it would appear that the obliteration, or rather obscuration, of the transverse striation of a fibre is due to the relationship of the individual fibrils in lateral apposition being disturbed, and the interfibrillary spaces being crowded with highly refractile granules of fat. This is shown by Figure 6, where the section was treated with ether to remove the fat granules, and thus render the fibres clearer. Now this section, when loaded with granules of fat, would present a somewhat similar appearance to Figure 7,* which is a typical pathological specimen of "fatty degeneration." Of course the fibres in the muscle of salmon are very much coarser than those of human muscle.

Bogdanow (†), in an account of his research on the muscle fat in horse flesh, describes that, after a single Soxhlet extraction, most of the fat in the connective tissue disappears, while the muscle fibre treated with 1 per cent. osmic acid stains brown (though lighter than non-extracted fibres); this staining gets lighter with each extraction. He adduces this in support of the statement that there exist two fats in flesh. Muscle of fish is much more friable than horse flesh, at least sections treated with ether lose all fat, as shown in Figure 6; but with prolonged action of alcohol I find that all connective tissue fat (even intercellular fat) disappears, but the interfibrillary fat glanules are not so easily removed. But from this it does not necessarily follow that the second fat is derived from the muscle plasma, whereas any attempt to prove its production by the muscle cell, owing to its resemblance to milk fat, would amount to "begging the question."

Scientific scepticism is always a great help towards the establishment of a firm foundation of truth. So the position assumed here is, that the usual microscopical evidence on "fatty degeneration of muscle" cannot be depended upon. A line of demarcation between fatty degeneration and fatty infiltration can hardly be drawn with a steady hand. Many cases of so-called fatty degeneration are merely such interfibrillary infiltration as occurs in the salmon's muscle. Actual breaking down of proteid molecules may take place, but any such statement has to be substantiated by results of chemical examination.

DESCRIPTION OF FIGURES.

THE FIGURES ARE REPRODUCTIONS OF MICRO PHOTOGRAPHS.

Fig. 1.—Muscle of salmon leaving the sea, with fat-globules between fibrils.
Fig. 2.—T.S. muscle fibres of salmon from upper reaches of river with very little interfibrillar fat.
Fig. 3.—T.S. muscle fibres of salmon caught in the sea with abundance of fat in the fibres.
Fig. 4.—L.S. muscle of salmon some time in the river showing little fat between the fibres.
Fig. 5.—L.S. muscle of salmon caught in the sea with abundance of inter-fibrous fat.
Fig. 6.—L.S. muscle of salmon treated with ether to remove fat globules to **show** apparent disappearance of transverse striation.
Fig. 7.—L.S. human muscle in state of fatty degeneration.

* This specimen was kindly lent by Dr. Robert Muir, Pathological Laboratory, Edinburgh University.
† Pflüger's Archiv., Bd. LXV.

10.—THE NATURE OF THE PROTEIDS OF SALMON MUSCLE.

By FRANCIS D. BOYD, M.D., F.R.C.P. Ed.

Dr. Dunlop, in his study of the muscles of the salmon during its sojourn in fresh water, finds that the proteids undergo a marked diminution, being in part transferred to the growing ovary and testis, and in part used as a source of muscular energy (p. 120).

These observations, however, do not deal with the nature of the proteids in muscle, and they leave uninvestigated the question of which proteids undergo this diminution. Since in the salmon we have an animal undergoing a very prolonged fast, the subject seemed of interest in relation to the pathology of starvation, and in view of results obtained by previous observers in relation to this question. Thus the investigations of Tiegel (1), Buchart (2), and Salvioli (3) on the question of the blood proteids during starvation have yielded somewhat contradictory results, and do not enable us to arrive at any definite conclusion.

Before considering the changes which the proteids undergo, it was necessary to study the nature of the proteids which occur in the muscles of salmon.

This paper is thus divided into two sections—A. The nature of the proteids of salmon muscle. B. The changes which these proteids undergo during the sojourn of the fish in fresh water.

A. *The Nature of the Proteids of Salmon Muscle.*

1.—Soluble Proteids.

Method. (1) In examining the soluble proteids of the salmon muscle, a portion of about 40 grammes of the flesh was taken, and all the bone and visible fibrous tissue separated. The muscle was then minced, pounded in a mortar, and extracted with normal salt solution. Normal salt solution was used, as von Fürth (4) has shown that stronger salt solution possesses disadvantages, in that the proteids become altered under the influence of the salt. It was found that all the soluble proteid could be extracted by treating the muscle twice with normal saline solution. The mixture thus obtained was filtered under pressure, the filtrate got being a faintly opalescent fluid. The fluid gave all the proteid reactions. No precipitation occurred on the addition of 1 per cent. of a 33 per cent. solution of acetic acid.

A. PROTEIDS COAGULABLE BY HEAT.

An examination of the fluid extract gave the following result :—

1. Dialysis. A quantity of the extract was put to dialyse in running water. After 48 hours a copious precipitate was present This preci-

pitate was separated. When redissolved in normal salt solution it gave
the reactions of *musculin*. The filtrate after the separation of the
musculin was subjected to fractional heat coagulation. The following
results were obtained :—

38 deg. C.	- -	Very slight opalescence.
40 deg. C.	- -	Decided opalescence.
45 deg. C.	- -	Milky fluid (filtered clear).
47·5 deg. C.	- -	Faint cloud.
48 deg. C.	- -	Distinct cloud
53 deg. C.	- -	Precipitate (filtered clear).
55 deg. C.	- -	Faint cloud.
60 deg. C.	- -	Decided cloud.
62 deg. C.	- -	Precipitate (filtered clear).
64 deg. C.	- -	Faint cloud.
68 deg. C.	- -	Very slight precipitate.
Heated to 80 deg. C.		No further precipitate.

The solution thus contained :—

(*a*) A proteid coagulated at from 38 to 45 deg. C. This would
correspond to the fibrin-like modification of myosin—the *soluble
myosin-fibrin* described by Fürth, rapidly formed at ordinary
temperatures from the paramyosinogen, the formation having probably taken place during the filtration of the original extract, a process
which even under pressure is necessarily tedious.

(*b*) A proteid coagulated at 53 deg C.—*myosinogin.*

(*c*) A proteid coagulated at 62 deg. C.—*myoglobulin.*

(*d*) A proteid coagulated at 68 deg. C. present in very small amount.
It is possible that this proteid is an albumin derived from the blood.
From the method in which the salmon flesh was obtained, it was
impossible to wash all the blood out of the tissues by perfusion.

2. Fractional coagulation with ammonium sulphate. To the fluid
extract a saturated solution of ammonium sulphate was added in the
proportion of 2 parts of the fluid extract to 1·5 parts saturated solution
of ammonium sulphate. The mixture then contained about 23 per
cent. ammonium sulphate. A copious white precipitate formed. This
was thrown upon a filter paper and washed with 23 per cent. solution
of ammonium sulphate. On being treated with water the most of the
precipitate went into solution. The fluid thus obtained gave all the
reactions of proteid of the nature of a globulin. Precipitation with
heat occurred at about 50 deg. C. The fluid gave a precipitate with
nitric acid, the xanthoproteic reaction and the biurét reaction. The
proteid answered closely to musculin or the paramysinogein of Halliburton (5).

After treatment of the muscle extract by fractional coagulation with
ammonium sulphate, the filtrate was saturated with ammonium sulphate.
A copious precipitate formed. This was removed, and the filtrate
examined and found to be entirely free from proteid. The precipitate
was collected, washed with saturated ammonium sulphate solution, and
redissolved in normal salt solution. The solution was clear, of a faintly
golden yellow colour, neutral in reaction, and gave all the reactions of a
proteid solution.

Fractional heat coagulation was carried out—

(*a*). When the extract was rapidly prepared and quite fresh

55 deg. C.	- -	A faint cloud
60 deg. C.	- -	Marked cloud.
65 deg. C. } 68 deg. C. }	- -	Distinct precipitate.

H

(b) When the extract had been allowed to stand 24 hours at the ordinary temperature of the room while in the process of filtering.

40 deg. C.	- -	Faint opalescence.
45 deg. C.	- -	Decided opalescence.
46 deg. C.	- -	Precipitate (filtered).
48 deg. C.	- -	Faint opalescence.
50 deg. C.	- -	Faint opalescence.
56 deg. C.	- -	More opalescent.
60 deg. C.	- -	Opalescence marked.
64 deg. C.	- -	Faint precipitate separating.
72 deg. C.	- -	Decided precipitate (filtered).
80 deg. C.	- -	No further precipitate.

Here again may be noted the formation of the soluble myosin fibrin, not precipitable on partial saturation with ammonium sulphate, but precipitable along with the myosinogen on complete saturation with ammonium sulphate—the heat coagulation of the soluble myosin-fibrin taking place from 40 deg. C. to 46 deg. C., the myosinogen and myoglobulin precipitating at from 56 deg. C. to 64 deg. C. These temperature results slightly differ from the temperature results got by Fürth in the case of the muscle proteids of dogs and rabbits.

From numerous observations the myosinogen of salmon muscle appears to coagulate at 55 to 58 deg. C.

Myosinogen of salmon muscle does not appear to be precipitated by dialysis, thus differing from paramyosinogen. It is completely precipitated by heat, gives a precipitate with nitric acid. It does not precipitate on the addition of dilute acetic acid. It gives the xanthroproteic reaction and the biuret reaction.

Method. (2) The muscle in some cases was extracted with strong salt (10 per cent. NaCl.) solution. It was then found that if the extracted fluid was rendered acid with acetic acid, as when 1 per cent. of a 33 per cent. solution of acetic acid was added, the whole proteid in the extract became precipitated. Some of the extract was taken and acetic acid added. The precipitate was removed, dried, and estimated. Some of the precipitate obtained with acetic acid was treated with a solution of carbonate of soda, when part went into solution. Thus :—

Total precipitate obtained with acetic acid,		-	-	2·415 per cent.
Soluble in sodic carbonate solution,	-	-	-	0·665 ,,
Residue,	-	-	-	-	-	-	1·750 ,,

B.—Proteoses and Peptone.

Fischel and Muira (6) have described the presence of peptone in muscle, and Halliburton described a muscle albumose. Halliburton, in his later work, however, has altered his opinion and agrees with Whitfield (7), who shows that proteoses and peptone are not found in the muscles of warm-blooded animals. The same can be said of salmon muscle. In examining for proteoses and peptone the fresh muscle was extracted with 10 per cent. salt solution, and the extract filtered under pressure. The filtrate was saturated with ammonium sulphate while boiling. On cooling the mixture was filtered. The filtrate was entirely free from proteid, thus showing the absence of peptone. The residue after filtration was treated with water and filtered. The filtrate gave no proteid reaction, showing the absence of albumose. These observations were repeated on half a dozen different fish which were selected under varying life conditions from the estuaries and upper waters of the rivers, and at different times of the year.

C.—The Presence of Albumin.

Halliburton describes an albumin as present in muscle plasma. I have made a very large number of observations on the muscle of salmon, but have been unable to satisfy myself of its constant presence. If the extract of salmon muscle was treated with an equal volume of saturated ammonium sulphate solution and rendered slightly acid with acetic acid, as a rule the entire proteid present was thrown down. If any proteid remained in solution it was a mere trace. The presence of a mere trace of albumin might result from the imperfect washing away of the blood from the flesh, as from the method in which the flesh was obtained it was impossible to perfuse, and so remove all the blood.

D.—Nucleo-Albumin.

The presence of nucleo-albumin in muscle is a subject which has been a good deal discussed of late. Whitfield (7) expressly denies its presence, stating that myosin is not a nucleo-albumin, because it contains no appreciable quantity of phosphorus in its molecule, because on gastric digestion only an insignificant residue is obtained which contains no phosphorus, and because when injected into the circulation it does not produce intravascular coagulation. He concludes that muscle contains no nucleo-albumin, because after digestion it yields only an insufficient residue, and this contains no appreciable quantity of phosphorus. In a still more recent paper Peckelharing (9), however, reaffirms the presence of nucleo-albumin in muscle.

I have made a large number of observations on the muscle of salmon. In all of them the phosphorus estimation was carried out by Dr. Noël Paton.

In extracting the proteid different methods were used.

(1) The flesh was extracted with 10 per cent. salt solution. The extract obtained was filtered first through muslin and then through filter paper under pressure. A large quantity of the extract was taken and to it 1 per cent. of a 33 per cent. solution of acetic acid was added. An enormous precipitate was obtained in every case on the addition of the acetic acid.* This precipitate was collected, washed, dissolved in 1 per cent. solution of sodic carbonate, reprecipitated with acetic acid, collected, and redissolved in sodic carbonate solution. The purification was repeated three times. The solution was then injected into the veins of a brown rabbit by Dr. Paton with negative results. It should be noted, however, that since this observation was made Halliburton (12) has pointed out that if nucleo-albumin be subjected to repeated purifications, it loses its power of producing intra-vascular coagulation.

(2) The precipitate obtained on the addition of acetic acid to the extract made with strong salt solution was separated, washed with acidulated water to free it of inorganic phosphates, and dried upon a porous plate. A small quantity, less than half a gramme, was examined for phosphorus, and a distinct but small trace was found.

(3) Several specimens of salmon muscle were examined by Peckelharing's method. Five hundred grammes of the salmon muscle was extracted with weak salt solution (1·5 grammes NaCl. per litre). The extract was filtered and the residue again extracted. The fluid thus obtained was filtered, and dilute acetic acid solution added ; a slight precipitate formed. Hydrochloric acid and liquor pepticus were added and the mixture put to digest for 48 hours. A considerable increase in the precipitate took place. It was thrown upon a weighed ash-free

* It should be noted that the acetic acid solution carried down the entire proteid in solution in most cases.

filter paper, washed with water, alcohol, and ether, dried, weighed, and examined for phosphorus. A number of observations were made. Whenever a **sufficient amount of** muscle (about 500 grammes) was **used,** phosphorus **was** found to **be** present. From this it may be concluded that salmon flesh does contain a soluble proteid, in the molecule of which an appreciable quantity of phosphorus is present. From the **behaviour of** the **extract** obtained with strong salt solutions, I would be **inclined to hold that the** soluble proteid was present as a globulin **loosely combined with a small** quantity of phosphoric acid either **as a true nuclein or a pseudo-nuclein.** From the results obtained I must **agree with Peckelharing, in so far as** salmon muscle is concerned, **that there is a nucleo-albumin present in the flesh.** It is possible **that** Whitfield's results were got by using small quantities of flesh. I myself got the **same results when dealing with an insufficient** bulk of muscle. These **results are confirmed by the observations on the** distribution of phosphrous **in the** muscles (p. 145).

II. Insoluble Proteids.

Solid Residue after Extraction.—The solid residue, after **extracting the soluble** proteids, was treated in the incubator for 24 hours **with a one per** cent. solution of caustic soda. The major part of the **residue went into** solution, the precipitate which remained representing **less than a third of the original material :** **this** consisted of collagen. **The solution gave a copious** precipitate when rendered faintly acid and **treated with alcohol.** It **gave reactions of. a** proteid solution. Some **of the alkaline fluid was rendered faintly acid, then a** copious precipitate **formed.** Liquor pepticus **and acetic acid** were added, when most of the **precipitate** disappeared. The mixture was digested for 48 hours, when **an increase in** the precipitate took place. This precipitate was separated **and** examined by Dr. Paton, when a distinct, though small, amount of phosphorus could **be** demonstrated. This corresponds with Karajew's (8) results. He **found** that the muscle of animals contained a body which **he** termed *myostromin*—a complex phosphorus containing albuminous body soluble neither in water nor in neutral salt solutions.

THE CURD.

Specimens of curded salmon were examined **with a view to** ascertaining the **nature of** the curd.

The flesh **was** plunged into boiling water **for three minutes,** when the curd became fixed. The curd was then **removed from** the muscle and examined. It gave all the reactions of a proteid coagulum.

It was treated with ether **and** pounded **in a** mortar, the residue collected and suspended **in water.** The fluid **gave a** marked **xanthroproteic** reaction.

B.—Changes in the Soluble Proteids at different seasons of the year, and under *differing conditions.*

A good deal of work has **been** done on the amount of soluble proteid in the **muscle of** different warm-blooded animals. Demant (10) places the proportion very low : 0·455 per cent. in the pectoralis major of rabbits. Danilewsky (13), in the case of man, found 3·68 per cent.

Von König (11) gives the composition of the nitrogenous part of the salmon muscle as follows :—

Per cent. of N. containing Substances.	Albumin. Soluble Proteids.	Flesh Fibre.	Gelatin.
19·39	3·39	11·02	1·50

In examining the muscle of salmon a mixed specimen of the thick and thin was taken. It was minced, pounded, and rubbed up with an equal weight of common salt; water was then added to the amount that would give a 10 per cent. solution of salt. The extract was passed through muslin, the residue again rubbed up with salt, water added, and the mixture filtered through muslin. The muscle was extracted in this way till all the soluble proteid had been removed. It was found that, as a general rule, three extractions were sufficient to remove all the soluble proteid. The extract was then filtered under pressure. The proteids were then estimated by heat coagulation and the gravimetric method. The following Tables show the results :—

ESTUARY.

River.	No.	Date.		Soluble Proteids. Percentage.
				Grammes.
Tweed	4	June	1895	2·98
Tweed	—	June	1895	3·33
Tweed	2	June	1895	3·3
Montrose	24	September	1895	3·97
Helmsdale	20	May	1896	2·8
Helmsdale	25	May	1896	2·65
*Dee	72	October	1896	3·97
				Average, 3·285

*This fish was in a specially fine state of nutrition.

UPPER WATER.

Tweed	34	November	1895	2·1
Tweed	42	December	1895	2·1
Tweed	44	December	1895	2·23
Tweed	—	December	1895	2·38
	21	May	1896	2·33
Helmsdale	62	October	1896	2·33
Spey	78	November	1896	3·1
				Average, 2·367

The Tables show a considerable excess of soluble proteid in the fish taken at the mouth of the river as compared with the fish in the upper reaches of the stream. The fish passing up from the sea are in good condition, with a large amount of soluble proteid in the muscle. As the season advances, and the fish has been up the river for some time, the amount of soluble proteid in the muscle shows a very decided diminution. The muscle is relatively poor in soluble proteid. Danilewsky (13) has shown that in warm-blooded animals the muscles that have least work to do are richest in globulin, the globulin in active muscles being probably used up in doing their work. The same may be held in the case of the fish. It passes up the river with a large store of soluble proteid in its muscle. As time advances and its muscles are called upon to supply energy, not only does the total amount of its muscle greatly diminish, but there is a marked diminution in the proportional amount of soluble proteid in the muscle. This difference is considerable, amounting (as shown in the Tables) to 0·918 grammes per cent. of soluble proteid—a percentage loss of 27 per cent. A comparison of this with the results obtained by Dr. Dunlop (p. 124) indicates that probably the proteid lost from the muscle is derived from these soluble proteids.

There are two fish of special interest. The fish from the Dee (No. 72), the last in the first Table, was noted on arrival to be specially well nourished; and on estimating the total soluble proteids they were found to be in very large amount. In the case of a kelt which was examined (an ill-nourished, lean fish which had not been down to the sea to feed), the proportion of soluble proteids was unusually low.

CONCLUSIONS.

From the foregoing the following conclusions may be deduced :—

(1) The salmon muscle contains three soluble proteids—Musculin or Paramyosinogen, Myosinogen, and Myoglobulin.

(2) That soluble Myosin-fibrin is rapidly formed from Paramyosinogen at ordinary temperatures.

(3) That these proteids are of the nature of globulins.

(4) That salmon muscle contains no albumin.

(5) That salmon muscle contains no proteoses and no peptone.

(6) That salmon muscle contains an insoluble albuminous body which contains phosphorus—Myostromin, which is probably a globulin combined with a true or a pseudo-nuclein.

(7) That the curd of salmon muscle is a proteid body.

(8) That there is a marked diminution in the percentage of soluble proteids in salmon muscle in fish which have been in the river for some time.

(9) That the muscle of the kelt is very poor in soluble proteid.

REFERENCES TO LITERATURE.

1. Tiegel—Pflügers Archiv. Vol. XXXII., s. 278.

2. Buchart—Arch. für exper. Path. und Phar. Vol. XVI., s. 322.

3. Salvioli—Du Bois Reymonds Archiv. für Physiol. Suppliment Band. 1879, s. 273.

4. Von Fürth—Arch. für exper. Path. und Phar. Bd. 36. 1895, s. 235.

5. Halliburton—Journal of Physiology. Vol. VIII. 1887, p. 188.

6. Fischel—Zeitschr. für physiol. Chemie. Vol. X., s. 14. Muira Virchow's Archiv. Vol. CI., s. 316.

7. Whitfield—Journal of Physiology. Vol. XVI., p. 487.

8. Karajew—Wratsch. 1895, No. 39, s. 1083.

9. Peckelharing—Zeitschr. für phys. Chemie. Bd. XXII., Heft 3, s. 245.

10. Damont—Zeitschr. für physiol. Chemie. Bd. III. 1879, s. 241.

11. Von König—Die Mensch. Nahrungs und Genusmittel. 2nd Part, s. 179.

12. Halliburton—Journal of Physiology. Vol. XXVII., p. 135.

13. Danilewsky—Zeitschr. für physiol. Chemie. Bd. V. 1881. s. 158.

11.—THE CHANGES IN THE AMOUNT OF PROTEID IN THE MUSCALATURE AND GENITALIA OF SALMON IN FRESH WATER.

By JAMES C. DUNLOP, M.D., F.R.C.P. Ed.

A study of the changes which the proteids of salmon undergo during the time of its ascending rivers and during its stay in fresh water is of interest, not only to those concerned with the marketable value of salmon as a food stuff, but also to the physiologist. In salmon we have an example of an animal which is required, during a long period of starvation, to do a large amount of muscular work, and at the same time to supply a large amount of substance for the requirements of rapidly developing genitalia. Proteid is certainly required for the growth of the genitalia, and almost as certainly for the production of some of the energy for muscular work, and for the necessary repair of the energy-yielding mechanism, the muscle; and so the question arises as to whether the salmon has in itself a sufficiency of proteids to supply these wants when it leaves the sea. Does a comparison of the fish after their stay in fresh water with fish captured in the estuary as they come from the sea warrant the conclusion that they have been obtaining the proteid required for these purposes from their own tissues?

Should there exist a store of proteid in fresh run salmon, it is natural to look for it in the musculature, which is so rich in proteid, and which forms such a large amount of the total weight of the fish. Consequently in this research the proteid of muscle has been fully examined. Of the great requirements for proteids mentioned above, one only, the growth of the genitalia, lends itself to measurement, and consequently it is this which is here specially considered, and any proteid consumption not accounted for by it is put down to the other great requirements, the supply of muscular energy and repair of the energy-producing mechanism.

A point of physiology, on which this study has a direct bearing, is how far the proteids of one tissue can be called on to supply the wants of another. If this examination of salmon shows that the proteid wants of the genitalia are supplied from the proteids of muscle, it indicates a very large transference from one tissue to another.

The subject has been previously studied by Miescher Ruesch (Internationale Fisherei-Austellung zu Berlin, 1880). He concluded that a fresh run salmon has an ample store of proteid to meet all its requirements in fresh water. The objections to Miescher's work have been considered in another portion of this Report (page 80). He gives percentage analyses of the proteids in a series of fish, but he made a definite quantitative analysis of the total amount of muscle proteid in only two fish, which were both caught in August. He did not make such estimations of fish in the later months with developed genitalia, but from other observations he argues, that these two fish should have lost a certain quantity of proteid from their muscle, and have gained a certain amount in their genitalia, and from this calculation he strikes a balance. But the second factor of this comparison, the condition of the later fish, being more or less hypothetical, cannot be considered sufficiently definite; his balance can only be accepted as a probable but not as an absolutely proved one.

Thanks to an ample supply of material from the Fishery Board, this investigation has been much fuller than Miescher's. A series of fish from the upper and lower waters, at different seasons, have been analysed and compared by the method already described in the introductory section of this Report. The results obtained enable a fairly accurate balance to be struck.

The method adopted for estimating the proteid of the organs was to determine the nitrogen contained, and from it to calculate the amount of proteid. This method being generally accepted as the most accurate available, does not require consideration. It gives the maximum possible amount of proteid, but its results are invariably too high, there always being present in tissues some extractive nitrogen. Results are stated as the amount of nitrogen, and changes in the amount of nitrogen are assumed to indicate changes in the amount of proteid.

In this section of the Report, when using the term proteid, not only are the true proteids referred to, but also all albuminoid substances. No attempt has been made in this part of the investigation to separate the various proteids and albuminoids from each other. A study of some of the individual proteids will be found in Dr. Boyd's section of the Report (page 112).

The nitrogen estimations were made by Kjeldahl's method on the dry residue left after the fats were extracted by ether (*vide* page 93). Knowing the amount of nitrogen in the dry residue, the amount of dry residue in the muscle or ovaries, and the total quantity of muscle or ovaries, the total amount of nitrogen in the muscle or ovaries of the fish is readily calculated. Thus, for an example :—

Fish 36.—Ovary weighs 126 grms., the dried residue of this after the extraction of fat was 29·3 per cent., and was found to contain 12·98 per cent. nitrogen. The ovary consequently contained $\frac{12\cdot98 \times 29\cdot3}{100} = 3\cdot8$ per cent. of nitrogen, and total amount of nitrogen in the ovary was $\frac{3\cdot8 \times 126}{100} = 4\cdot13$ grms.

In the following Tables will be found the results of the analyses. In them the fish have been classified in the same manner as in the earlier sections of this work. The first group considered are the female fish of 1896, next the male fish of 1896, then the fish caught in 1895, and lastly kelts. The 1896 females are taken up first, because they form much the largest group.

TABLE 1.

Showing Nitrogen of Female Estuary Fish, 1896.

MAY AND JUNE.

No.	River	Length	Weight	Muscle — Thick			Muscle — Thin			Muscle — Total		Ovaries			Remarks
				Weight	N.%	N.Total	Weight	N.%	N.Total	Weight	N.Total	Weight	N.%	Total	
15	Spey	73	3650	1755	3·05	58·85	585	3·09	18·07	2340	76·2	49	2·58	1·08	*Exceptionally poor.
16	Dee	75	3860	1898	3·80	72·04	652	3·18	20·09	2550	92·1	60	3·01	1·2	*{ Exceptionally poor, gall
17	Spey	79	4295	2033	3·80	67·08	677	3·12	21·12	2710	88·2	43	2·57	1·1	bladder empty.
20	Helm	73	3365	1962	3·25	64·42	654	2·91	19·0	2620	83·1	56	3·16	1·76	
25	Helm	74	4095	1995	3·29	65·63	665	3·02	19·08	2660	84·7	70	3·24	2·3	+{ Exceptionally poor, old
27	Dee	73	3795	1833	3·25	59·57	610	3·05	18·63	2444	78·2	57	3·35	1·9	wounds on left side.
29	Dee	68	3020	1520	3·27	48·16	440	2·88	12·67	1760	59·9	47	2·44	1·14	

JULY AND AUGUST.

No.	River	Length	Weight	Weight	N.%	N.Total	Weight	N.%	N.Total	Weight	N.Total	Weight	N.%	Total	Remarks
36	Dee	78	4752	2206	3·75	84·47	768	3·31	25·42	3974	109·9	126	3·80	4·76	
40	Dee	71	3810	1881	3·31	54·87	627	2·95	26·28	2508	71·1	39	3·06	1·19	
43	Spey	75	4225	1985	3·02	59·94	661	2·9	19·16	2646	79·1	66	2·98	1·96	*Exceptionally heavy fish.
46	Helm	71	4445	2268	3·19	72·17	756	3·09	23·48	3021	102·6	73	3·33	2·42	
51	Spey	74	4306	2190	3·56	74·52	710	3·18	20·09	2940	97·1	80	3·85	2·68	
55	Dee	81	3675	2700	3·96	109·9	920	3·06	28·15	3680	137·2	158	3·62	5·71	

OCTOBER AND NOVEMBER.

No.	River	Length	Weight	Weight	N.%	N.Total	Weight	N.%	N.Total	Weight	N.Total	Weight	N.%	Total	Remarks
65	Spey	87	8520	3600	2·93	101·96	1200	2·93	35·16	4800	137·1	1025	4·07	41·76	*Winter Salmon.
73	Dee	90	8184	3296	3·43	111	1078	3·62	39·02	4311	150·0	1160	4·12	47·79	*Winter Salmon.
74	Dee	91	8134	3416	3·52	120·1	1138	3·20	36·42	4554	130·5	990	3·26	32·27	
29	Helm	88	7415	3758	3·33	125·1	1252	1·9	23·79	5010	148·9	43	2·12	0·91	
72	Dee	89	7214	3368	3·10	111·4	1197	2·14	25·60	4700	136·9	70	2·6	1·82	

*Excluded from calculation of average.

TABLE II.

SHOWING NITROGEN OF FEMALE UPPER WATER FISH, 1896.

No.	River	Length	Weight	Muscle — Thick			Muscle — Thin			Total		Ovaries			Remarks
				Weight	N. %	N. Total	Weight	N. %	N. Total	Weight	N. Total	Weight	N. %	N. Total	
MAY AND JUNE.															
11	Spey	69	2880	1895	2·13	29·98	965	1·95	7·71	1860	37·3	34	2·72	0·92	
12	Spey	73	3740	1766	3·44	69·73	588	3·08	21·63	2354	82·4	57	2·78	1·62	
21	Helm	76	4065	2918	3·41	68·81	672	3·26	21·84	2690	99·6	97	2·93	2·84	
28	Helm	73	4167	2925	3·4	63·85	675	3·11	29·96	2700	89·8	111	3·15	3·96	
31	Spey	75	3550	1659	3·28	44·41	553	2·96	16·47	2212	70·9	122	2·61	3·23	
32	Dee	85	5447	2468	3·32	51·83	822	3·35	27·58	3290	108·5	255	3·46	8·82	
JULY AND AUGUST.															
37	Spey		4010	1523	3·20	58·33	607	2·89	17·54	2430	75·9	161	3·89	5·94	
42	Spey		3501	1620	3·51	66·86	540	3·13	16·93	2160	78·8	214	4·11	8·79	
43	Helm	66	2572	1149	3·22	36·99	383	3·97	11·75	1582	48·7	103	3·53	4·61	
47	Dee	74	3990	1740	3·4	49·93	480			2520		258	3·90	9·43	Analysis lost.
49	Helm	70	3490	1623	3·34	54·29	541	3·93	16·39	2164	70·4	160	3·79	6·06	
OCTOBER AND NOVEMBER.															
62	Helm	74	3700	1241	2·89	35·4	114	2·63	11·63	1658	56·	822	4·91	12·91	
66	Helm	74	3490	1128	2·82	31·8	576	2·76	10·57	1501	12·2	750	4·24	3·48	
67	Dee	69	3270	1118	2·96	33·43	352	2·87	10·67	1490	41·1	691	4·90	23·84	
65	Helm	73	3845	1258	2·54	32·38	112	2·29	9·35	1650	44·7	511	4·68	26·34	
68	Helm	64	2575	908	2·70	24·51	302	2·54	7·87	1210	32·2	655	4·63	25·71	
69	Helm	73	3990	1298	2·82	34·06	362	2·45	8·54	1610	35·9	1109	3·65	40·44	
70	Dee	66	2875	855	2·88	24·62	265	2·73	7·78	1140	32·1	360	3·63	22·99	

*Excluded from calculation of average.

From the foregoing Tables 1 and 2, the following shorter Tables are constructed. In them the total nitrogen of the muscle and of the ovaries is expressed, not as the total of the individual fish, but for purposes of comparison as the total of the fish of standard length (*vide* page 6).

TABLE III.

Showing Amount of Nitrogen per Fish of Standard Length in Muscle and Ovaries.

FEMALE ESTUARY FISH, 1896.

May and June.

No.	Muscle.			Ovaries.	
	Thick. N per Cent.	Thin. N per Cent.	N. Total per Fish of Stndrd Length.	N. per Cent.	N. per Fish of Stndrd Length.
16	3·80	3·18	218	3·01	2·85
20	3·25	2·91	219	3·16	4·55
25	3·29	3·02	209	3·24	5·50
27	3·25	3·05	201	3·35	4·92
Average,	3·40	3·04	210	3·19	4·45

July and August.

36	3·75	3·31	211	3·80	10·10
40	3·31	2·95	225	3·06	3·03
45	3·02	2·90	188	2·98	4·60
51	3·56	3·18	238	3·35	6·50
55	3·96	3·06	228	3·62	10·70
Average,	3·52	3·08	218	3·36	7·

October and November.

65	2·93	2·93	213	4·07	63·5
73	3·43	3·48	206	4·12	65·5
74	3·52	3·44	208	3·26	42·7
Average,	3·29	3·28	209	3·82	57·2

TABLE IV.

Showing Amount of Nitrogen per Fish of Standard Length in Muscle and Ovaries.

FEMALE UPPER WATER FISH, 1896.

May and June.

No.	Muscle.			Ovaries.	
	Thick.	Thin.	N. Total per Fish of Stndrd Length.	N. per Cent.	N. Total per Fish of **Stndrd** Length.
12	3·44	3·68	195	2·85	3·85
21	3·41	3·25	206	2·93	6·47
31	3·28	2·98	174	2·64	7·95
32	3·32	3·35	178	3·46	14·36
Average, . .	3·38	3·31	188	2·97	8·16

June and July.

No.					
37	3·20	2·89	165	3·69	12·8
42	3·50	3·13	197	4·11	23·4
43	3·22	3·07	172	3·55	12·7
49	3·34	3·03	205	3·79	17·4
Average, . .	3·32	3·03	185	3·78	16·6

October and November.

No.					
62	2·89	2·65	123	4·01	81·4
63	2·82	2·76	104	4·14	76·6
64	2·99	2·87	134	4·30	90·7
66	2·54	2·20	107	4·33	94·0
67	2·70	2·54	103	4·05	81·8
69	2·82	2·45	108	2·68	99·7
70	2·88	2·73	113	2·83	78·5
Average, .	2·81	2·6	113	4·05	86·1

In considering the observations shown in the foregoing Tables, I shall first take up the condition of the muscle as shown by the percentage of nitrogen contained; then the total amount of muscle proteid in the fish of standard length as shown by the amount of nitrogen; then consider

the ovaries in a similar manner ; and lastly, consider the inferences that
may be drawn from the changes observed.

1. Percentage of Proteid in Muscle.—Two samples of muscle from each
fish were examined, one from the under part of the fish (thin), the other
from the dorsal part (thick). It will be observed in the tables that the
"thick" muscle is, at all seasons, and in both estuary and upper-water fish,
richer in nitrogen than the "thin" muscle. This difference in the amount
of nitrogen, and consequently in the amount of proteid, emphasises the
importance of analysing both portions of the musculature as already
noted (p. 80). The smaller percentage of proteid in the thin muscle is
probably dependent on the larger percentage of fat there (*vide* p. 83),
the fat of course increasing the weight reduces the percentage of other
constituents present.

The amount of proteid of the muscle in estuary fish is fairly constant
throughout the season. It is slightly higher in July and August than
in May and June, and slightly lower in October and November than in
the earlier months. The fish which show the greatest divergence from
the average are Nos. 72 and 79. These fish were caught in the estuary
late in the year (winter salmon), had undeveloped ovaries, and would not
spawn till the following year. Their muscles were laden with fat, and
consequently the percentage of proteid was lower than the average.

The muscles of the upper-water fish show more change. Those of the
fish caught in October and November have a much smaller amount of
proteid than the earlier fish. This percentage diminution is not due to
increase of fat, as in 72 and 79, the amount of fat being diminished, but
is due to an increase in the water of the muscle (page 84). The upper-
water fish from May to August have a fairly constant percentage of
proteid, and that percentage is nearly the same as in estuary fish
throughout the series.

TABLE V.

Showing the percentage of nitrogen in the muscle of the different
groups of fish :—

	Estuary.		Upper Water.	
	Thick.	Thin.	Thick.	Thin.
May and June - - -	3·40	3·04	3·38	3·31
July and August - - -	3·52	3·08	3·32	3·03
October and November - -	3·29	3·28	2·81	2·60

II. Total Muscle Proteid of the Fish.—In considering this point two
comparisons suggest themselves. Firstly, comparing the muscles of the
fish of the later periods of the year with the earlier ones; and,
secondly, comparing the muscles of the upper-water fish with those of
the estuary fish.

The estuary fish, as is seen in Table III, have a nearly constant
amount of muscle nitrogen throughout the season, the average for the

three periods being 210, 218, and 209 grammes of nitrogen in the fish of standard length, the average for the year being 213. Some exceptions to this were observed, notably Fish 15, 17, 29 (*vide* Table 1), the nitrogen of these amounting to only 196, 177, and 177 grammes per fish of standard length. Nos. 29 and 17 were exceptional in other ways. No. 29 had wounds on its side, No. 17 had an empty gall bladder, but about No. 15 nothing abnormal is noted.

The upper-water fish show marked changes in the amount of muscle nitrogen, the amount decreasing as the season advances, the decrease being much more marked in the later part of the season. The amount of nitrogen per fish of standard length was in May and June 188 grms., in July and August 185 grms., in October and November only 113 grms.

The large fall of the amount of muscle nitrogen in October and November will be considered after the gain of nitrogen by the ovaries has been discussed.

A comparison of the upper-water fish with those of the estuary shows that all through the year the upper-water fish have less muscle nitrogen than the estuary fish, the deficit being greatest in October and November.

III. Percentage of Proteids in the Ovaries.—An examination of the figures given in Tables 1 and 2 shows that there is throughout the season a steadily increasing percentage of proteid matter in the ovaries, both in estuary and in upper-water fish. In estuary fish the amount of nitrogen in the ovaries rises from 3·19 per cent in May and June to 3·36 per cent in July and August, and 3·82 per cent in October and November. In the upper-water fish the corresponding figures are 2·97 per cent, 3·78 per cent, and 4·05 per cent.

IV. Total Proteids of the Ovaries.—This also shows a steady increase during the season; not only are the ovaries richer in nitrogen as the season advances, as stated above, but they are increasing in weight the whole time. The increase of proteid takes place both in estuary and in upper-water fish, but it is greater in the upper water fish. These changes are shown in the following table.

TABLE VI.

Showing amount of ovarian nitrogen per cent. and in fish of standard length:—

	May and June.		July and August.		Oct. and Nov.	
	Estuary.	Upper Water.	Estuary.	Upper Water.	Estuary.	Upper Water.
Nitrogen per cent.	3·19	2·97	3·36	3·78	3·82	4·05
Nitrogen per fish of standard length .	4·45	8·16	7·0	16·6	57·2	86·1

Comparing this Table with that referring to the total amount of muscle nitrogen during the three periods, it is evident that it is during the time of the greatest ovarian increase that the muscle loses most.

A comparison between the loss of proteid from muscle and the gain of proteid by ovaries, when the salmon is in fresh water, is shown in the following table, the fish being compared with each other in the manner adopted by Dr. Paton (p. 81).

TABLE VII.

Showing balance of nitrogen between muscle and ovaries per fish of standard length:—

Date and Source of Fish.	Average Muscle Nitrogen.	Loss from Muscle.	Average Ovary Nitrogen.	Gain by Ovaries.	Surplus Loss from Muscle available for Energy, &c.
May to August— Estuary, .	214		5·7		
July to August— Upper Water,	185	29	12·4	6·7	22
May to August— Estuary, . .	214		5·7		·
October to November— Upper Water, .	113	101	86·0	80·3	20

These balances clearly show that the amount of proteid lost from the muscle is ample to supply the wants of the growing ovaries, as in each case the loss from the muscle is greater than the gain by the ovaries. As previously stated, the surplus loss may be considered as showing that there is a consumption of proteid to meet the other requirements—the production of energy and the repair of energy-producing mechanism. It will be seen that this surplus loss amounts in each instance to a very considerable quantity—in one case to 22 grammes of nitrogen per fish of standard length, in the other to 20 grammes.

Twenty-two grammes of nitrogen are equivalent of 137·5 grammes of proteid, the energy value of which is 560 large calories or 240,000 kilogramme-metres, an amount of energy sufficient to raise a fish of standard length weighing 10 kilogrammes to a height of 24,000 metres (or 78,000 feet). Similar calculation shows the other surplus loss has an energy value of 217,000 kilogramme-metres, and is sufficient to raise a fish of standard length to a height of 21,700 metres (or 71,000 feet).

V. Conclusions.—The inferences to be drawn from this examination of the female fish of 1896 may be summed up as follows:—

1. Estuary fish have more muscle proteid than upper-water fish.

2. The amount of muscle proteid in the upper-water fish diminishes as the season advances. The October and November upper-water fish are very poor in muscle proteid, having little more than half the amount of muscle proteid that estuary fish have.

3. The ovaries of both estuary and upper-water fish gain proteid as the season advances.

4. The estuary fish and the upper-water fish of the early months have sufficient proteid in the musculature to supply the wants of the growing ovaries, and from the deficit of muscle proteid found in the late upper-water fish, it is probable that there is a transference of proteid from the musculature to the ovaries.

5. The deficit of muscle proteid in upper water fish is so large that after allowing for the requirements of the ovaries there remains a surplus loss. This surplus loss is available for the liberation of a large amount of energy.

MALE FISH, 1896.

The examination of the male fish received during 1896 corroborates the results obtained from the examination of the female fish of that year. Though the number received and examined was somewhat small, it was sufficiently large to show that the male fish are affected by their stay in fresh water in exactly the same way as the female fish.

The results of the observations are shown in the following Tables :—

TABLE VIII.

SHOWING NITROGEN OF MALE ESTUARY FISH, 1896.

No.	River.	Length.	Weight.	Muscle. Thick. Weight	Thick. N.%	Thick. N.Total	Thin. Weight	Thin. N.%	Thin. N.Total	Total. Weight	Total. N.Total	Testes. Weight	Testes. N.%	Testes. N.Total	Remarks.
							MAY AND JUNE.								
19	Annan	80	4445	2153	3·2	689	717	2·82	291	2870	89·	x	3·29	·28	
							JULY AND AUGUST.								
20	Spey	74	5280	2675	3·82	88·7	593	2·88	25·7	3564	114·4	14	1·87	·96	
21	Spey	57	7010	3453	3·39	117·	1151	3·04	35·	4604	152·1	21	2·26	·52	
22	Annan	84	5650	2483	3·79	89·6	807	2·29	18·	3290	117·6	45	2·58	1·16	
							OCTOBER AND NOVEMBER.								
23	Dee	108	12769	5708	3·22	183·8	1902	2·86	54·4	7610	238·2	270	4·16	11·23	
24	Dee	48	2890	1245	3·33	41·4	415	3·03	22·6	1660	61·	96	4·02	3·86	

TABLE IX.

SHOWING NITROGEN OF MALE UPPER-WATER FISH, 1896.

MAY AND JUNE.

No.	River.	Length.	Weight.	Muscle.								Testes.			Remarks.
				Thick.			Thin.			Total.		Weight.	N. %	N. Total.	
				Weight.	N. %	N. Total.	Weight.	N. %	N. Total.	Weight.	N. Total.				
34	Dee	68	2650	1211	3·33	40·5	413	3·01	12·4	1654	52·8	14	2·13	·30	

JULY AND AUGUST.

No.	River.	Length.	Weight.	Thick Weight.	Thick N.%	Thick N.Total	Thin Weight.	Thin N.%	Thin N.Total	Total Weight.	Total N.Total	Testes Weight.	Testes N.%	Testes N.Total
38	Helm	74	3570	1841	3·50	64·4	613	3·31	29·3	2454	84·7	10	1·91	·19
39	Dee	77	3835	1815	3·26	59·2	605	2·85	17·2	2420	76·4	44	1·88	·83
54	Dee	79	3880	1796	3·65	63·4	578	3·31	19·1	2314	82·5	101	2·32	2·34

OCTOBER AND NOVEMBER.

No.	River.	Length.	Weight.	Thick Weight.	Thick N.%	Thick N.Total	Thin Weight.	Thin N.%	Thin N.Total	Total Weight.	Total N.Total	Testes Weight.	Testes N.%	Testes N.Total
68	Dee	74	3280	1284	2·78	35·7	428	2·47	10·6	1712	46·3	109	3·43	3·74

TABLE X.

SHOWING PER FISH OF STANDARD LENGTH THE AMOUNT OF NITROGEN
IN MUSCLE AND TESTES IN MALE ESTUARY FISH, 1896.

MAY AND JUNE.

No.	Muscle.			Testes.	
	Thick. N. %	Thin. N. %	N. per Fish of Standard Length.	N. %	N. per Fish of Standard Length.
19	3·2	2·82	174	3·29	0·51
Average,	3·2	2·82	174	3·29	0·51

JULY AND AUGUST.

No.	Muscle.			Testes.	
56	3·32	2·88	222	1·87	0·37
59	3·39	3·04	231	2·16	0·78
61	3·70	2·23	181	2·58	1·95
Average,	3·47	2·72	211	2·20	1·03

OCTOBER AND NOVEMBER.

No.	Muscle.			Testes.	
71	3·22	2·86	209	4·16	8·84
75	3·33	3·03	172	4·02	12·27
Average,	3·28	2·95	190	4·09	10·55

TABLE XI.

SHOWING PER FISH OF STANDARD LENGTH THE AMOUNT OF NITROGEN
IN MUSCLE AND TESTES IN MALE UPPER-WATER FISH, 1896.

MAY AND JUNE.

No.	Muscle.			Testes.	
	Thick. N. %	Thin. N. %	N. per Fish of Standard Length.	N. %	N. per Fish of Standard Length.
34	3·26	3·01	168	2·13	0·94
Average,	3·26	3·01	168	2·13	0·94

JULY AND AUGUST.

No.	Muscle.			Testes.	
38	3·50	3·31	206	1·91	·85
39	3·26	2·85	167	1·88	1·81
54	3·65	3·31	167	2·32	4·75
Average,	3·47	3·16	180	2·04	2·47

TABLE XI.—*Continued.*

OCTOBER AND NOVEMBER.

68	2·78	2·47	114	3·43	9·23
Average,	2·78	2·47	114	3·43	9·23

These Tables show the following facts about the male salmon :—

1. In all the groups of fish the thick muscle contains more **proteid** than the **thin**.

2. The **upper-water fish** have a smaller **percentage of nitrogen** in their muscle than the estuary fish.

3. The total amount of **proteid** in the musculature is less in upper-water fish than in estuary fish, the upper-water fish of October and November being especially poor.

4. The **testes** become richer in nitrogen as the season advances.

5. The **total** proteid of the testes is greater as the season advances.

6. The **loss** of proteid from the muscle is greater than the gain by the testes, there being a surplus loss available for the production of energy and processes of repair.

This surplus loss is much greater than in female fish, and consequently a much larger amount is available for energy.

The changes being exactly comparable to those observed in female salmon, it may be concluded that **male salmon**, like **female salmon**, have sufficient proteid stored in their muscles to meet all their requirements in fresh water, and that this proteid is called on to supply the wants of the growing **testes**, and for the liberation of energy.

SALMON OF 1895.

The small number of fish available for analysis prevents averages being struck and accurate comparisons being made. The results obtained, and to be seen in Tables 12, 13, 14, and 15 show that the amounts of proteid in the muscle and genitalia of these fish vary within the same limits as were found in the 1896 fish. No notes are available as to whether the fish included in these Tables are fresh run or have been in fresh water for some time, but by comparing the amount of muscle nitrogen per fish of standard length with that of the 1896 fish they can be readily classified. Thus 31, 19, and 40 are evidently fresh run, as their muscle contains as much nitrogen as the muscle of the fresh run 1896 fish of the corresponding periods, and 29, 39, and 43 are evidently fish which have been in fresh water for some time, their musculature being in the condition found in upper-water fish of 1896 of the corresponding periods. The condition of the genitalia corroborates this contention.

TABLE XII.

Showing Nitrogen of Female Fish, 1895.

ESTUARY.

MAY AND JUNE.

No.	River	Length	Weight	Muscle — Thick			Muscle — Thin			Total		Ovaries			Remarks
				Weight	N. %	N.Total	Weight	N. %	N.Total	N. %	N.Total	Weight	N. %	N.Total	
2	Tweed	—	—	—	3·52	—	—	2·67	—	—	—	71	2·52	1·78	

JULY AND AUGUST.

No.	River	Length	Weight	Weight	N. %	N.Total	Weight	N. %	N.Total	N. %	N.Total	Weight	N. %	N.Total	Remarks
8	N. Esk	76	4077	—	4·05		64	3·12	177			90	1·28	1·28	Weight of Muscle unknown.
10	N. Esk	76	9130	1943	3·50		647	2·71	250			218	3·73	8·13	Weight of Muscle unknown.
23	N. Esk	78	6786	—	3·54			3·26	817			236	1·41	3·33	

OCTOBER, NOVEMBER, AND DECEMBER.

No.	River	Length	Weight	Weight	N. %	N.Total	Weight	N. %	N.Total	N. %	N.Total	Weight	N. %	N.Total	Remarks
27	N. Esk	94	2491	—	2·77	139·4	1400	2·64	40·2	179·6	5690	920	2·72	11·62	Grilse.
31	N. Esk	76	9200	4160	3·35	53·6	500	2·86	10·2	63·8	2680	965	1·02	40·2	
32	N. Esk	91	4335	2180	2·46	74·4	980	2·95	24·8	99·2	8740	817	1·66	12·96	
39	N. Esk		7686	2810	2·63			2·67				—	Lost.		

TABLE XIII.

SHOWING NITROGEN OF MALE FISH, 1895.

ESTUARY.

No.	River.	Length.	Weight.	Muscle — Thick. Weight.	N.%	N. Total.	Thin. Weight.	N.%	N. Total.	Total. N.%	N. Total.	Testes. Weight.	N.%	N. Total.	Remarks.
MAY AND JUNE.															
4	N. Esk	...	...	...	3·17	...	...	3·61	...	...	...	7·7	2·62	·15	Weight of Muscle unknown.
5	N. Esk	...	...	...	3·3	...	...	3·25	...	...	...	39·	2·09	·31	Weight of Muscle unknown.
6	N. Esk	...	...	...	3·42	...	...	3·1	...	...	...	8·	3·12	·25	Grilse, Weight of Muscle unknown.
JULY AND AUGUST.															
19	N. Esk	68	3440	1829	3·23	59·0	420	2·83	11·9	22·8	70·9	11·1	1·48	0·16	
OCTOBER, NOVEMBER, AND DECEMBER.															
40	N. Esk	97	11256	4630	3·10	144·1	1550	2·88	38·9	6260	181	398	2·81	8·65	Grilse.
43	N. Esk	73	8280	1344	2·41	32·8	448	2·55	11·4	1780	44·2	111·5	2·44	1·8	

TABLE XIV.

SHOWING AMOUNT OF NITROGEN PER FISH OF STANDARD LENGTH IN MUSCLE AND OVARIES OF FEMALE FISH, 1895.

No.	Muscle			Ovaries	
	Thick. N. %.	Thin. N. %.	Total per Fish of Standard Length.	N. %.	N. per Fish of Standard Length.
MAY AND JUNE.					
2	3·52	2·67	–	2·52	–
JULY AND AUGUST.					
8	4·05	3·12	–	1·27	–
10	3·55	2·74	197	3·73	18·5
23	3·64	3·26	–	1·41	–
OCTOBER, NOVEMBER, AND DECEMBER.					
27	2·77	2·64	–	2·72	–
31	3·35	2·86	215	4·02	48·3
29	2·46	2·05	144	4·09	29·5
39	2·65	2·67	114	Lost.	–

TABLE XV.

SHOWING PER FISH OF STANDARD LENGTH THE AMOUNT OF NITROGEN IN MALE FISH, 1895.

No.	Muscle			Testes.	
	Thick. N. %.	Thin. N. %.	N. per Fish of Standard Length.	N. %.	N. per Fish of Standard Length.
MAY AND JUNE.					
4	3·17	3·60	–	2·02	–
5	3·00	3·25	–	2·09	–
6	3·49	3·10	–	3·12	–
JULY AND AUGUST.					
19	3·23	2·83	224	1·48	1·0
OCTOBER AND NOVEMBER.					
40	3·10	2·38	198	2.81	9·5
43	2·41	2·55	104	2·44	5·6

KELTS.

In the following Tables will be seen the results obtained from the analyses of kelts. One was received in 1895, the others early in 1897.

TABLE XVI.

SHOWING NITROGEN OF KELTS, 1895 AND 1897.

No. and Year	River	Length	Weight	Muscle						Total		Ovaries			Remarks
				Thick			Thin								
				Weight	N. %.	N. Total.	Weight	N. %.	N. Total.	Weight	N. Total.	Weight	N. %.	N. Total.	
14. 1895	—	—	—	—	3·18	—	—	2·93	—	—	—	—	1·87	—	
80. 1896	Spey	87	5431	2303	2·95	67·9	767	2·84	21·8	3070	89·7	29	1·62	·47	
81. 1896	Spey	93	6847	3000	2·91	87	1000	2·72	27·2	4000	114·2	52	1·70	·88	
82. 1896	Spey	92	6235	2625	2·89	76·1	875	2·77	24·2	3500	100·3	65	1·67	1·08	
83. 1896	Spey	99	4900	1969	2·76	54·3	680	2·54	25·7	2480	66·0	38	1·39	·53	

TABLE XVII.

SHOWING NITROGEN PER CENT. AND PER FISH OF STANDARD LENGTH
IN KELTS, 1895 AND 1896.

No.	Muscle.				Ovaries.	
	Thick. N. %	Thin. N. %	Total N. per Fish of Standard Length.		N. %	Total N. per Fish of Standard Length.
14	3·18	2·93			1·87	
80	2·95	2·84	135		1·62	·71
81	2·91	2·72	142		1·70	·88
82	2·90	2·77	138		1·67	1·35
83	2·76	2·54	96		1·39	·9

TABLE XVIII.

COMPARING KELTS WITH UNSPAWNED FISH.

		Unspawned Fish.			Kelt.
		Estuary.	Upper Water.		
			July and August.	October & November.	
Muscle,	Thick, N. %,	3·40	3·32	2·81	2·94
	Thin, N. %. .	3·13	3·03	2·60	2·74
	N. per Fish of Standard Length	} 212	185	113	128
Ovaries,	N. %, . .	3·46	3·78	4·05	1·64
	N. per Fish of Standard Length	} 4·45—57·2	16·6	86·1	0·9

A comparison of these results with those of the female unspawned
fish of 1896 shows the following :—

1. The muscle of the kelt contains a smaller percentage of proteid
than all the unspawned fish except those from the upper waters late in
year.

2. The total amount of muscle nitrogen per fish of standard length
is less than that of the unspawned fish, the same late upper-water
unspawned fish being excepted.

3. The percentage of proteid in the ovary is less than in all unspawned
fish.

4. The total nitrogen per fish of standard length in the ovary is less
than in all unspawned fish.

It will be seen in the Tables that out of the four fully analysed kelts
three have more nitrogen in their muscle than the upper-water fish of the
later months of the year. There are two explanations of this, the one
that they represent not the late upper-water fish, but the late estuary

fish; the other, **that there** has been a retransference **of** proteid from
the ovary to the muscle. Both these are feasible, but the former is
probable, from the fact that the size of the fish corresponds, not with
the **upper-water fish** of October **and** November, **but** with **the estuary
fish of** these **months. The fourth examined kelt has so little muscle**
nitrogen, only **96 grms. per fish of standard length, that it might
be** considered **to represent a** late upper-water **fish; the other three,**
averaging **138 grms., might** represent **the late** estuary **fish. On the
other** hand, **the loss from the** ovaries in and **after** spawning is **so much**
greater than **the gain by the** muscle, **that it is** possible that **some of**
this proteid **may have been** transferred; **but, as no measure of the loss
in ova is available, it cannot** be stated **how much ovarian proteid
remained to be retransferred to** the muscles.

THE CARBOHYDRATES OF THE SALMON.

NOTE BY D. NOEL PATON, M.D.

In the **present** investigations the Carbohydrates have **not been**
studied. They **are very** briefly considered by Miescher (Histochemischen
and Physiologischen Arbeiten 1897, **Bd. II., s. 325).** He finds that
sugar is **present in the** blood and liver **and glycogen in the** muscles and
liver, though in small amounts, even **when the salmon has been** long in
the river.

12.—THE FATS AND PROTEIDS STORED IN THE MUSCLE CONSIDERED AS A SOURCE OF MUSCULAR ENERGY.

By D. NOËL PATON, M.D. F.R.C.P.Ed.

The foregoing study of the changes in the fats and proteids affords a basis for the elucidation of the part played by each of these in the liberation of energy available for the muscular work of the fish.

The fats and proteids lost from the muscles during the sojourn of the fish in fresh water, over and above those accumulated in the ovaries and testes, must be decomposed for the liberation of energy.

This matter is of interest, firstly, because it shows whether the energy evolved from the fats and proteids decomposed is adequate for the needs of fish during its stay in fresh water ; and, secondly, because it elucidates the question, not hitherto studied by physiologists, of the source of the energy in cold-blooded animals.

The following Table shows the amount of energy in work units—Kilogrammètres—set free from each of these substances throughout the season. In these Tables the fats and the nitrogen are given in grms. :—

FEMALE SALMON 1896.

Fat Balance per Fish of Standard Length.

	Muscle.	Ovaries.
Estuary (May to August),	770	15
Upper Water (July and August)	478	46
	Loss—292	Gain—31

Available for liberation of energy, 261 grms. = 1,025,730 Kgms.

To this should be added about 20 grms. of fat from the pyloric appendages ; and about 20 grms. of fat from the liver (see p. 100), say 40 : which makes 300 grms., or 1,179,000 kgms.

	Muscle.	Ovaries.
Estuary (May to August),	770	15
Upper Water (Oct. and Nov.),	159	204
	Loss—611	Gain—189

Available for liberation of energy, 422 grms. = 1,658,460 Kgms.

Nitrogen Balance per Fish of Standard Length.

	Muscle.	Ovaries.
Estuary (May to August),	214	5·7
Upper Water (July and August)	185	12·4
	Loss— 29	Gain— 6·7

Available **for liberation of energy, 22 grms. = 240,000 Kgms.**

	Muscle.	Ovaries.
Estuary (May to August), .	214	5·7
Upper Water (Oct. and Nov.).	113	86·0
	Loss— 101	Gain—80·3

Available for liberation of energy, 20 grms. = 217,000 Kgms.

MALE SALMON 1896.

The number of male fish analysed is **too small** to allow of any good **average** being struck as regards the amount of material available for the evolution of energy. Especially is this the case as regards the later part of the season, since only one male **fish from the** upper waters was obtained in October and November.

In the earlier part of the season the **evolution of energy appears** about the same in the male as in the female.

TABLE

Fat Balance per Fish of Standard Length.

	Muscle.	Testes.
Estuary (May to August), .	756	0·74
Upper Water (July and **August)**	428	2·41
	Loss—328	Gain— 1·67

Available for liberation **of** energy, 326 grms. = 1,271,400 Kgms.

	Muscle.	Testes.
Estuary (May to August), .	756	0·74
Upper Water (Oct. and Nov·),	103	6·34
	Loss—653	Gain—5·60

Available for liberation of energy, 647 grms. = 2,523,300 Kgms.

Nitrogen Balance per Fish of Standard Length.

	Muscle.	Testis.
Estuary (May to August),	192	0·77
Upper Water (July and August)	180	2·47
	Loss— 12	Gain—1·70

Available for liberation of energy, 10 grms. = 103,000 Kgms.

	Muscle.	Testis.
Estuary (May to August)	192	0·77
Upper Water (Oct. and Nov.),	114	9·23
	Loss —78	Gain—8·46

Available for liberation of energy, 70 grms. = 763,000 Kgms.

FEMALE FISH.

To August—   Kgms.

    Energy liberated from fats, . . . 1,025,730
    Energy liberated from proteids, . . 240,000
    Energy from proteids to energy from fats, as 1 : 4·2.*

To November—   Kgms.

    Energy liberated from fats, . . 1,658,460
    Energy liberated from proteids, . . 217,000
    Energy from proteids to energy from fats, as 1 : 7·6.

MALE FISH.

To August   Kgms.

    Energy liberated from fats, . . . 1,271,400
    Energy liberated from proteids, . . 109,000
    Energy from proteids to energy from fats, as 1 : 11·6

These Tables bring out several points of very great interest.

1. Of the fats lost from the muscle of *female* fish to August, only a very small moiety—12 per cent.—goes to the ovaries, the remaining 88 per cent. is available as a source of energy. Taking the metabolism to November, 30 per cent. of the fats go to the ovary and 70 to energy

2. Of the proteids lost from the muscle to July and August in the female, 23 per cent. are transferred to the ovaries, 77 per cent. are available for energy; but later in the season the proteid lost from the muscle is almost entirely transferred to and built up in the growing ovaries, little or none being available for muscular energy.

3. It thus follows that while in the earlier months the energy of muscular work is derived from the fats to the extent of 81 per cent., and from proteids to the extent of 19 per cent., during the later months the fats are almost the sole source of energy.

---

*If the internal fats are also considered the proportion will be 1 to 4·9.

4. In the *male*, of the fats lost from the muscle to August, 5 per cent. are accumulated in the testes, while 95 per cent. are available for energy. Of the proteids 14 per cent. go to the testes, leaving 86 per cent. as a source of energy.

5. In the period to August, when male and female fish can be compared, the energy liberated per fish of standard length was, in the female, equivalent to 1,265,000 Kgms.,* while in the male it was equivalent to 1,380,000 Kgms. In the female, where fat accumulation in the ovaries is large, a greater proportion of energy appears to be derived from the proteids of the muscle than in the male, where the testes is comparatively poor in fats. Here the fats of the muscle yield a larger proportion of energy than in the female. In the female to August, of the total available energy, about 20 per cent. is derived from the proteids, while in the male only 9 per cent. is obtained from this source.

In ascending the river the salmon has not only to raise its weight to a given height, but it has to overcome the friction of the stream.

The former factor in the expenditure of energy is easily calculated if the elevation of the upper waters of the river is known.

The following Table gives the elevation of the upper waters of the three rivers, and the work done in raising a fish of standard length and average weight—10 Kgs.

	Height in Metres.	Work Done in Kgms.	Energy Evolved from Fats and Proteids to August in Female Fish
Dee - -	225	2250	
Spey -	216	2160	1,266,000
Average,	220	2205	1,266,000

The energy employed in merely lifting the weight of the fish is to the whole energy expended as 1 to 570.

An enormous surplus of energy is thus available for the work of overcoming friction. To form anything like a correct idea of what this may be is at present impossible, for information as regards the rate of progress of salmon is wanting, and the work done by the fish will necessarily vary with the rate at which the upward progress is made.

But these figures show that for this work an enormous surplus of energy is available from the combustion of the fats and proteids which disappear from the muscles throughout the sojourn of the fish in fresh water.

Pflüger has recently (*Pflüg. Arch. Bd., XLVI.*) most strenuously maintained the view that the proteids are the great source of the energy for muscular work. The present observations very clearly prove that, *in a cold-blooded animal, fats are a source of energy, and that they play a much more important part than the proteids.*

---

* The Kilogrammètre is the work done in lifting a Kilogramme through one metre. It is equivalent to 7·24 foot-pounds.

13.—THE PHOSPHORUS COMPOUNDS OF THE MUSCLE AND GENITALIA OF THE SALMON, AND THEIR EXCHANGES.

By D. NOËL PATON, M.D. F.R.C.P. Ed.

Phosphorus forms an essential constituent of every living creature, and in the higher animals it is very widely distributed in the different tissues in various combinations. In the bones and other tissues it occurs as inorganic *phosphates*, possibly in loose organic combination. It also occurs as stored nutrient material in the yolk of the egg in at least two forms—(*a*) as *pseudo-nucleins*, substances yielding a proteid and phosphoric acid when split, and (*b*) as *lecithin*—in which a molecule of the fatty acid radicle of a glycerine fat is replaced by phosphoric acid linked to a peculiar nitrogenous base, cholin. Lastly, in protoplasm, and more especially in the nucleus of the cells, it occurs in *nuclein* compounds—compounds in which a proteid is linked to nucleic acid—a substance which, on being broken down, yields phosphoric acid and various nitrogenous bases such as xanthin.

The most complex of these phosphorus compounds are the true nucleoproteids. It is these compounds which, in the nuclei of cells, play so important a part in the various phenomena of living matter.

That lecithin is a forerunner of these substances is shown by the study of the disappearance of lecithin in the egg during incubation as the embryo develops.

Whether the pseudo-nucleins are also forerunners of these true nucleins has, so far as I am aware, never been considered. The fact that they exist in large quantities in the yolk of the egg would indicate that in them the phosphorus is also stored for use in the construction of nucleins, while the possible combination of their phosphorus in thymic acid—a derivative of nucleic acid—seems a further indication of their relations to the true nucleins.

It would be difficult to find a more suitable object for the study of some of the transformations of phosphorus than the salmon during its prolonged fast, when its genitalia are being built up from its other tissues.

Miescher Ruesch does not deal fully with the question. He states (p. 183) that the ovarian fluid, which readily exudes from the ripe ovary when broken down, contains no less than 20 per cent. of lecithin, besides a nuclein the amount and characters of which he does not discuss. He points out that this has "general interest because both these substances, especially the last named, are contained only in very small amounts in muscle. The formation of the ovarian fluid can thus only take place by the muscle substance being taken up from the flesh and laid on in the

144 *Investigations on the Life-History*

ovaries, but out of the albumen and fat and phosphorus-containing salts of the muscle the characteristic combination of the egg must be formed by the profoundest chemical changes."

He further states that "there is no doubt that the trunk muscles contain more than·enough phosphoric acid to yield the phosphorus of the ovary, with 1·1 per cent. $P_2 O_5$ in the fresh ovary, as against 2·3 per cent. to 2·6 per cent. in the dry substance of the trunk muscles." He gives no further figure in support of his statement.

It must be remembered that in the bones of the fish there is an abundant supply of phosphorus, and that this, as well as the supply in the muscles, may be available for the ovaries.

Before discussing the exchanges which take place between the phosphorus of the muscle and the genitalia, it will be well, in the first instance, to consider in a general manner the nature of the phosphorus compounds in each.

I. Nature of Phosphorus Compounds.

Methods.

The muscle after preservation in alcohol was pounded and repeatedly extracted with hot alcohol and subsequently with ether. The extract was evaporated and the phosphorus determined by igniting with caustic potash and nitre (both phosphorus—free) dissolving in water, acidifying, filtering through paper extracted with acid, and precipitating the phosphates with ammonia-magnesia mixture, igniting, and weighing the ash. From the magnesium phosphate the lecithin was calculated by multiplying by 7·00. This factor was selected because it is midway between the factors for stearin lecithin and olein lecithin. The phosphorus was determined by multiplying the magnesium phosphate by 0·279.

The phosphorus not as lecithin was determined as follows :—From 1 grm. to 2·5 grms. of the residue after extraction with ether were taken for analysis. This was burned with caustic potash and nitre and dissolved in water. Strong nitric acid was added, and the solution was heated on a steam bath for several hours. Next day nitrate of ammonia and molybdate of ammonia in excess were added, and the solution was kept warm and allowed to stand for 12 or 24 hours. It was then filtered, and the filtrate again treated with molybdate of ammonia, and if any precipitate fell this was also thrown on the filter paper. The yellow precipitate was washed through the paper with ammonia solution, and the phosphates precipitated with ammonia - magnesia mixture and ignited. From the magnesium phosphates the phosphorus was calculated.

To ascertain the phosphorus as nucleins and as phosphates, the residue after extraction of lecithin was then extracted with ·2 per cent. HCl, and the phosphorus determined in the extract. This extract contained hardly any organic matter. .

The phosphorus of the residue was finally determined.

In this way the amount of phosphorus as lecithin, as inorganic phosphates, and as nucleins or pseudo-nucleins was ascertained.

A.—Phosphorus Compounds of the Muscles.

The amount of phosphorus in various forms in the muscles of different animals has already been investigated by Katz. (Pflüg. Arch., Bd. 63, p. 1-18.)

In the eel and pike—the two fish examined by him—he obtained the following results :—

Phosphorus in 100 Parts Dried Muscle.

	Total.	In Watery Extract. Phosphates.	In Alcohol Extract (Lecithin).	In Residue (Nucleins).
Eel, - -	0·48	0·40	0·055	0·03
Pike, - -	1·03	0·83	0·075	0·12

These analyses indicate that by far the greatest amount of phosphorus is held in the muscles as phosphates. The amount in lecithin and nucleins is comparatively small and unimportant.

The distribution of phosphorus was investigated in two salmon, one from the estuary early in the year, and one, also from the estuary, later in the year—Nos. 14 and 76.

TABLE I.

Distribution of Phosphorus in Muscles.

		Ether Extract.	Water Extract.	Residue.	Total.
14	Thick, . .	0·042	0·131	0·056	0·228
	Thin, . .	0·046	0·094	0·041	0·181
76	Thick, . .	0·060	0·095	0·055	0·210
	Thin, . .	0·060	0·119	0·063	0·242

This Table shows that in the salmon, as in other fish, the greater part of the phosphorus of the muscle is in the form of inorganic phosphates. The amount contained in lecithin appears larger in the salmon than in the pike and eel, and the proportion as nucleins is also larger.

A very different distribution of this element is found in the ovaries and testes.

B. PHOSPHORUS COMPOUNDS OF THE OVARIES.

1. General Distribution of Phosphorus. The distribution of phosphorus in the ovary was determined in the same way as in the muscle in Nos. 14 and 76.

TABLE II.

		Ether Extract.	Water Extract.	Residue.	Total.
14	.	0·114	0·057	0·114	0·285
76		0·150	0·075	0·189	0·414

K

In the ovaries the amount of phosphorus as inorganic phosphates is very small. The amount as lecithin nearly equals the amount as nucleins, or pseudo-nucleins, and in these two together the proportion is about four times that in the inorganic phosphates.

2. *Nature of Organic Phosphorus Compounds.* The ripe ovary of a November salmon was passed through the mincer to break up the ova, and strained through muslin. The viscous brown-yellow fluid was then diluted to three times its volume with solution of chloride of sodium, and repeatedly extracted with ether.

A. ETHER EXTRACT.

On evaporation, this separated into (I.) a rich orange-red solution; (II.) a copious white gelatinous precipitate.

1. *The orange-red solution*, on standing, threw down fine needle-like crystals readily redissolved on heating. The ether was distilled off and the residue dried at 100° C.

a. Of this 3·042 grm. were saponified by Kossel and Obermüller's method and the fatty acids extracted. These weighed 2·929 grm., *i.e.*, 92·8 per cent. of the ether extract.

b. In 2·066 grm. the presence or absence of phosphorus was determined in the usual way. There was no precipitate with the ammonia-magnesia mixture, so lecithin is absent.

c. 4·903 grm. were saponified and the cholesterin determined in the ether filtrate.

$$\text{Cholesterin} = 0·012 \text{ grm.}$$
$$= 0·24 \text{ per cent. of the ether extract.}$$

The clear ether extract consists almost entirely of ordinary fats with a little cholesterin and some lipochromes.

II. *The White Precipitate from the Ether Extract.* This separated out as the ether evaporated, and was only redissolved with difficulty in a large excess of ether. A small quantity was dried at 110°C. It weighed 1·225 grm. In this phosphorus was determined in the usual way.

$$\text{Magnesium Phosphate} = 0·103 \text{ grm.}$$
$$= 0·0287 \text{ grm. Phosphorus.}$$
$$= 2·3 \text{ per cent. of Phosphorus.}$$

Lecithin contains about 4 per cent. *The white precipitate seems to be largely composed of lecithin.*

Thus, by simple extraction with ether, a large quantity of fats, lecithin, some cholesterin, and pigment are removed from the ovarian fluid.

B. PROTEID OF OVARY.

The fluid left, after extraction with ether, had a dirty brown colour, and was somewhat viscous.

On pouring it into water a stringy white precipitate with the proteid reactions at once fell. This was washed with water and redissolved in salt solution.

1. *Lecithin is combined with the Proteid.*

A quantity of the proteid precipitated with alcohol was repeatedly extracted with boiling absolute alcohol, and the extract filtered hot and evaporated to dryness. A brownish sticky residue resulted. This was dried at 100° C., and weighed 0·248 grm. In this the phosphorus was determined in the usual way.

Magnesium Phosphate = ·037 grm.
= ·259 grm. Lecithin.

After extraction of the fats with ether, lecithin is left in closer combination with the proteids.

2. Presence of Phosphorus in Proteid.

After extraction with hot alcohol and ether, and drying at 100° C., 1·134 grm. of the proteid were used for determination of phosphorus.

Magnesium Phosphate = ·031 grm.
= ·0086 grm. Phosphorus.
= ·74 per cent. of Phosphorus.

After removal of the lecithin the proteid contains a further amount of phosphorus.

3. *Reactions of Proteid of Ovary.* The proteid was found to be insoluble in water; soluble in dilute KHO, Na_2CO_3, and in neutral salts. On the addition of an acid it is precipitated, and is soluble only in a considerable excess of acetic acid, but in a small excess of mineral acids, *e.g.*, ·2 per cent. HCl. In a weak solution of neutral salts it is not precipitated on boiling; but in a saturated solution of $NaCl$, in which it is soluble, a copious precipitate is thrown down on heating. It is precipitated by a half-saturated solution of $(NH_4)_2SO_4$. The filtrate gives a cloud on boiling and the proteid reactions—so a trace of an albumin is present. A stream of CO_2 gives no precipitate. After being precipitated for some time the proteid loses its solubility in neutral salts.

4. *Does the Proteid contain a Nuclein or Pseudo-Nuclein?* 2·26 grm. were digested for 20 hours in artificial gastric juice. The residue on drying weighed 0·109 grm.

A large quantity of the proteid was dissolved in ·2 per cent. HCl and liquor pepticus added. On digesting a copious precipitate was thrown down. This precipitate was washed with hot alcohol and ether, and the phosphorus determined in ·626 grm.; it contained 0·0237 grms. P., or 3·7 per cent.

A considerable quantity of the moist proteid was boiled in 5 per cent. H_2SO_4 for several hours. A brownish solution resulted, with a slight brown precipitate. This was filtered off. The filtrate was treated with baryta water to neutralisation; the baryta was separated by a stream of CO_2; the fluid was then filtered and ammonia added. No precipitate formed. $Ag\,NO_3$ was then added, and no precipitate resulted. Hence *nuclein bases are absent.* The brownish solution, when rendered alkaline and boiled with Fehling's solution caused no reduction.

5. *The Proteid contains Iron. (Note by E. D. W. Greig, M.B., C.M.)* In the earlier analyses of the ovarian proteid the iron was not determined.

In an elaborate investigation Walter (Zeitschrift für physiol. Chemie, Bd. XV., 1891, p. 477) showed that in the ovary of the carp both the proteid and the nuclein derived from it contained iron.

I have also determined the quantity of iron present in the pure ovarian proteid, and also in the nuclein obtained from it by artificial digestion.

The *proteid* was obtained in the following manner:—The fresh ovary was passed through a mincer, and then strained through muslin. It was then repeatedly extracted with ether, and about four times its volume of salt solution (strong) added. This was then poured into a large quantity of distilled water and repeatedly washed by decantation. It was finally repeatedly extracted with hot alcohol and ether. The powder thus obtained was dried and weighed.

The iron analysis was carried out in exactly the same manner as that which is fully described in my investigation on "The Exchange of Iron between the Muscle and Ovary of Salmon" (p. 156). In 2 grms. of the ovarian proteid, ·45 mgms. Fe, = 0·022 per cent. Fe. was found.

The *Nuclein* was obtained by digesting the pure proteid in 2 per cent. HCl and liquor pepticus at 40°C. The residue was extracted with hot alcohol and ether, and washed with water till acid-free. A very fine light powder was left, which was dried at 110°C. and weighed. The iron analysis was conducted as before. Result :—2·5 gr. ovarian nuclein contained 1·6 mgms. Fe, = 0.064 per cent. Fe.

The percentages are lower than those obtained by Walter.

Ovarian Proteid, -   -   = 0·117 per cent. (Walter).
„   Nuclein, -   -   = 0·252 per cent.   „

This lower result is probably due to the greater accuracy of the method here employed, though it may be due to difference in the composition of the ichthulin from the different fish.

6. Nature of the Ovarian Proteid.—The proteid of the ovary is in very close combination with a certain amount of lecithin, and when this is removed a pseudo-nuclein is left. This proteid resembles closely the ichthulin somewhat imperfectly described in 1854 by Valenciennes and Frémy (Comptes rendus, T. 38, 1854, p. 471), and more fully investigated by Walter (Ztsch. f. phys. Chem., Bd. XV., p. 477, 1891) in the ovary of the carp, though having a somewhat higher proportion of phosphorus and a smaller proportion of iron. Neumeister is probably right in concluding (Physiol. Chem.) that it is not a chemical individual but a mixture of a proteid with lecithin on the one hand and a pseudo-nuclein on the other.*

C.—Phosphorus Compounds of the Testes.

The characteristic phosphorus compounds of the testes have already been so ably and exhaustively investigated by Miescher-Ruesch (Arch. f. exp. Path. u. Pharmac., Bd., 37, p. 100, 1896) that it is unnecessary here to deal with them at length. The nuclein present contains true nucleic acid linked to a base which Miescher has called protamin. In this paper, which has been most admirably put together by Professor Schmeideberg after the death of the author, the characters of the protamin and of the nucleic acid are carefully described. The quantitative composition of the seminal fluid is discussed, the fluid part, the heads, and the tails of the spermatozoa having been isolated and separately studied. It is shown that while the tails contain an albuminous substance with fats and lecithin, and are very poor in other phosphorus-containing substances (p. 142), the heads are rich in nucleic acid and protamin, and very poor in fats and lecithin. The following (p. 139) gives the composition of the heads by one method :—

Nucleic acid, - - - - -	60·50 per cent.
Protamin extracted by HCl, -	19·78 „
Other substances „ - -	2·94 „
Still unknown residue, - - -	16·78 „
	100,00 „

---

*Since this was written a paper by Miescher bearing upon these questions has been published, "Histochemischen und Physiologischen Arbeiten von Freidrich Miescher B II.

After further investigation (p. 148), he concludes that, after extraction with alcohol and ether, the heads of the spermatozoa contain

    60·50 per cent. of Nucleic acid.
    35·56   ,,   Protamin.
  or 96·06   ,,   Neutral nucleate of protamin.

Some observations on the composition of the unripe testes are also given. Miescher devised an ingenious method of dissolving away the cell protoplasm, and he was thus able to study the chemistry of the nuclei from which the heads of the spermatozoa are developed. He shows that these nuclei contain nucleic acid and a proteid, " nuclear albuminose," but almost certainly no protamin It is considered possible that the albuminose is the precursor of the protamin. From his observations on the metabolism of the salmon, he concludes that the necessary phosphorus is carried to the testes as lecithin and there stored in the protoplasm, which becomes the tail of the spermatozoon. He further points out that since both nucleic acid and protamin are richer in nitrogen than proteids, a very great loss of these from the muscles must be necessary to yield the nitrogen, while a considerable nitrogen-free part must be liberated which will be available as a source of energy. The essential nature of the phosphorus compounds in the testes are thus very different from the compounds of the ovaries.

To determine the distribution of phosphorus in these structures the testes of 53 and 68 were analysed in the same manner as the ovaries.

TABLE III.

No.	Ether Extract.	Water Extract.	Residue.	Total.
53	·063	·068	·161	·292
68	·063	·040	·178	·278

The percentage amount of phosphorus in the testes is thus no larger than that in the ovaries. The amount in the form of lecithin is markedly less, that in inorganic phosphates about the same, while the phosphorus in the nucleins is in somewhat larger amount.

2. EXCHANGE OF PHOSPHORUS BETWEEN MUSCLE AND OVARIES.

The first question which it appeared desirable to investigate is :—

(*a*) *Is the lecithin stored in the trunk muscles sufficient to yield the lecithin which accumulates in the ovaries?*

Table IV. gives the per cent. of lecithin in ovaries and muscles in four typical fish from the estuary from May to August ; and in three typical fish from the upper waters in October and November.

[TABLE.

TABLE IV.
Percentage of Lecithin in Ovaries and Muscle.
ESTUARY FISH.

No.	Ovaries.	Thick.	Thin.
16	*2·10	0·76	0·93
27	2·35	0·70	0,68
40	1·99	0·56	0·48
45	2 23	0·70	0·53
Average,	2·17	0·68	0·65

UPPER-WATER FISH.

62	2·36	0·59	0·74
64	4·09	0·49	0 94
67	3·18	0·42	0·56
Average,	3·21	0·46	0·74

Table V gives the lecithin per Standard Fish.

TABLE V.
Lecithin per Standard Fish in Ovaries and Muscle.
ESTUARY FISH.

No.	Ovaries.	Muscle.
16	1·99	48·0
27	3·45	43·6
40	2·17	37·8
45	3·62	41·2
Average, . .	2·81	42·6

UPPER WATER FISH.

60	47·8	25·6
64	86·8	27·1
67	64·2	17·5
Average, . .	66·1	23·4

Gain of ovaries, 63·3 grms.   Loss of muscle, 19·2 grms.

From this Table it is seen that the gain of lecithin in the ovary is enormously in excess of the loss of lecithin from the muscle. The muscle per unit of length lost only 19·2 grms., while the ovary gained no less than 63·3 grms.

The source of the fatty acids of the ovarian lecithin presents no diffi-

---
* It has already been shown (p. 146) that the extraction of lecithin by ether from the ovaries is probably not complete. In the ovary there is thus even a larger quantity than is indicated by these figures.

culty. It has already been shown that the muscle contains an ample supply of fats to meet all such requirements.

It is the source of the phosphorus which requires investigation. The insufficiency of the lecithin phosphorus of the muscle is shown in the Table VI.

TABLE VI.

Phosphorus in Lecithin per Standard Fish.

ESTUARY FISH.

No.	Ovaries.	Muscle.
13	0·079	1·92
27	0·138	1·74
40	0·087	1·51
45	0·145	1·65
Average, . .	0·109	1·70

UPPER-WATER FISH.		
60	1·91	1·02
64	3·47	1·08
67	2·57	0·70
Average, . .	2·97	0·93

Gain of ovaries, 2·861. Loss of muscle, 0·877.

About 2 grms. of phosphorus in the lecithin of the ovary in the fish of standard length is thus not derived from the phosphorus of the lecithin of the muscle.

(b) *Is the phosphorus in other combinations in the muscles sufficient to yield the phosphorus of the lecithin and other phosphorus-containing bodies in the ovaries?*

To determine this point, three typical fish from the estuaries from May to August, and three typical fish from the upper waters in October and November, were selected.

TABLE VII.

Percentage of Phosphorus not in Lecithin in Muscle and Ovaries

ESTUARY FISH.

No.	Ovaries.	Thick.	Thin.
27	0·275	0·270	0·208
40	0·238	0·292	0·249
45	0·233	0·280	0·243

UPPER-WATER FISH.			
64	0·333	0·250	0·208
67	0·435	0·209	0·218
70	0·363	0·199	0·176

TABLE VIII.

Phosphorus not in Lecithin per Standard Fish in Muscle and Ovaries.
ESTUARY FISH.

No.	Ovaries.	Muscle.
27	0·403	16·0
40	0·259	20·6
45	0·363	17·0
Average, . .	0·342	17·8

UPPER-WATER FISH.

64	7·03	10·9
67	8·79	8·1
70	7·34	6·9
Average, . .	7·72	8·6

Ovaries gained 7·478 grms.   Muscle lost 9·200 grms.

Taking together the phosphorus as lecithin and the phosphorus not as lecithin—

Muscle loses 9·97 grms.   Ovaries gain 10·339 grms.

The phosphorus lost from the muscle is just about sufficient to yield the phosphorus laid on by the ovaries, provided no phosphorus is excreted or used in other ways.

(c) Phosphorus of Liver.

A possible source of the phosphorus is the liver. I have shown that in many animals a considerable storage of lecithin may occur in this organ. (J. of Phys., Vol. XIX., 1896, p. 167.)

Although the amount of fat in the liver of the salmon is comparatively small, it seemed desirable to ascertain if, during the period of fasting, loss of phosphorus from the liver goes on to any marked extent.

For this purpose six typical fish were selected, and the phosphorus of the liver as lecithin and as other compounds was determined in the usual manner.

Table IX. gives the results of these analyses.

TABLE IX.
Phosphorus of Liver.
ESTUARY FISH.

No.	Per Cent. of Phos. in Lecithin.	Per Cent. of Phos. not in Lecithin.	Total per Cent. of Phos.	Total Phos. per Fish of Standard Length.
27	0·051	0·053	0·104	0·173
40	0·072	0·016	0·088	0·167
45	0·065	0·026	0·091	0·195

TABLE IX.—*Continued.*

UPPER-WATER FISH.

No.	Per Cent. of Phos in Lecithin.	Per Cent. of Phos not in Lecithin.	Total per Cent. of Phos.	Total Phos. per Fish of Standard Length.
64	0·039	0·065	0·104	0·145
67	0·085	0·106	0·191	0·357
70	Analysis lost.	0·084		

These observations indicate that there is no great storage of phosphorus in the liver, and no marked diminution in the amount as the season advances. The liver cannot be regarded as the source of the phosphorus for the ovaries, and it is rather to the bones that we must look for any supply of this element over and above that yielded by the muscles which may be necessary for the growth of the ovaries.

Considering the conditions of the observations the balance between the phosphorus lost from the muscle and that gained by the ovaries is fairly close, and would seem to indicate that *the phosphorus stored in the muscle is sufficient to yield the phosphorus gained by the ovaries.*

It is not to be expected that the loss of phosphorus from the muscle should exceed that required by the ovaries, as is the case with the fats and proteids, for here no question of supply of energy is involved.

In its diet of marine fishes the salmon has an abundant supply of phosphorus, e.g., in the phosphates of the bones, and it is of interest to notice that in the intestinal mucus numerous opaque yellow nodules are to be seen, which are composed of crystals of carbonate of lime stained with bile pigments. Whether these are lime salts from which the phosphorus has been removed we have no information.

EXCHANGE OF PHOSPHORUS BETWEEN MUSCLE AND TESTES.

Of the phosphorus exchange in the male, Miescher-Ruesch writes as follows (*loc. cit.* p. 216):—"The ripening of the testes in the male does not require so large an amount of albumin and fat as in the ovary of the female; but instead of this, according to my investigations, just as much more phosphates for the construction of the various constituents of the spermatozoa rich in phosphorus. If we take the weight of the ripe testis at 5 per cent. of the weight of the body, with 25 per cent. dry solids with 11·3 per cent. phosphoric acid, we get 0·141 per cent. of the body weight in phosphoric acid which the growing testis must take from the blood, more than a half the amount contained in the ripe ovary of a female of the same size."

He does not in this paper further consider the source of this phosphorus, though, as pointed out on p. 148, he deals with this question in his later work.

To study the exchanges of phosphorus between muscle and testes, the phosphorus as lecithin, i.e. the phosphorus in the ether extract, and the phosphorus not as lecithin was determined in six typical male fish, three from the estuaries and three from the upper waters throughout the season.

TABLE X.

Per Cent. Phosphorus in Lecithin.

ESTUARY FISH.

No.	Thick.	Thin.	Testes.
13	0·028	0·026	0·022
56	Analysis lost.		
71	0·032	0·042	0·053

UPPER-WATER FISH.

No.	Thick.	Thin.	Testes.
34	0·035	0·036	0.045
39	Analysis lost.		
68	0·022	0·037	0·037

TABLE XI.

Per Cent. Phosphorus not as Lecithin.

ESTUARY FISH.

No.	Thick.	Thin.	Testes.
13	0·243	0·261	0·265
56	0·234	0·197	0·243
71	0·250	0·270	0·296

UPPER-WATER FISH.

No.	Thick.	Thin.	Testes.
34	0·252	0·228	0·318
39	0·188	0·204	0·243
68	0·193	0·200	0·240

It is thus seen that there is no great increase in the **per cent.** of phosphorus as the testes increase.

To show the extent of the exchange of phosphorus the amount of that substance per fish of standard length as lecithin and **not** as lecithin may be compared in 13, a fish leaving the sea in May, and in 68, a fish in the upper water in October.

Table XII. gives the Phosphorus **as Lecithin (A)**; the Phosphorus not as Lecithin (B); and the Total **Phosphorus (C)** per fish of standard length.

TABLE XII.

No.	MUSCLE.			TESTES.		
	A.	B.	C.	A.	B.	C.
13	1·43	13·90	15·33	0·002	0·029	0·031
68	1·32	10·27	11·59	0·103	0·645	0·748

Loss of Phosphorus by Muscle 3·740
Gain   ,,   Testes 0·717
_____

Surplus   3·023 grms.

It would thus appear that the store of phosphorus in the muscle is far more than sufficient to yield the phosphorus required in the constructive changes in the testes.

It is interesting in this connection to note that in the male during the summer months there is a great growth of bone in the snout, and it is highly probable that some of the phosphates stored in the muscles is utilised in this process.

GENERAL CONCLUSIONS.

These observations on the distribution and exchanges of phosphorus in the salmon throughout its sojourn in fresh water show that a supply of phosphorus, partly as inorganic phosphates, partly as lecithin, is stored in the muscles as these grow and become loaded with fat during the stay of the fish in the sea.

While the fish is in the river this stored phosphorus is transmitted from the muscle to the growing ovaries and testes, and in being transmitted undergoes changes in its chemical combinations. In the ovary the simple phosphates of the muscle are, (a) combined with fatty acids and cholin to form the abundant supply of lecithin; (b) combined with proteids to form the pseudo—or para—nuclein ichthulin—which is so abundant a constituent of the ovary. (c) In the testes, on the other hand, the phosphorus of the muscle phosphates is elaborated with the more complex nucleic acid and combined with the characteristic base—protamin.

The source of these two nitrogenous bases—the cholin of the lecithin and the protamin of the nuclein of the testes—we do not at present know. They must ultimately be derived from proteids since these are the only nitrogen-containing constituents of the body, and Miescher has suggested that the "albuminose" of the early testis is the forerunner of the protamin of the ripe organ.

There is no evidence that the transference of phosphorus from the muscles to the genital organs is not direct. There is no evidence that the liver plays any intermediate part. In fact, all the evidence tells against such an idea. It is apparently in and by the active protoplasm of the growing ovaries and testes that these profound chemical changes are carried out.

These observations indicate that from the simple phosphates of the muscle, by synthesis, lecithin, the ichthulin of the ovaries, and the nucleic acid of the testes are all built up. That lecithin is a stage in the production of the ichthulin of the ovary and nuclein of the testis seems to be indicated first by its accumulation in the muscles, and second by its appearance in both ovaries and testes. That it is a forerunner of the nucleins and phosphorus compounds of the embryo is undoubted. As to the source of the phosphorus stored in the muscle, there can be little doubt that it is in great measure derived from the phosphates in the bones of herring and other fish upon which the salmon feeds. The evidence is against the idea that the lecithin of the muscles is also directly taken from the lecithin in the food, for Hasebrock (Ztsch. f. phys. Chem., Bd. XII., 150) has shown that in the intestine lecithin is split up into glycero-phosphoric acid, cholin, and fatty acids. There must thus be synthesis of this material in the body.

14.—THE EXCHANGE OF IRON BETWEEN MUSCLE AND OVARIES OF SALMON.

By E. D. W. GREIG, M.B. Edin.

It has already been shown in another part of this Report (p. 147) that iron is present in the paranuclein of the ovaries of salmon.

This investigation was undertaken to determine (1) whether the quantity of iron present in the ovaries increases during their growth, and if so (2) from what source it is derived. In considering the possible sources of supply, it was at once obvious that the food ingested could be excluded, as evidence has been adduced (p. 13 *et seq.*) to show that salmon do not take nourishment during their sojourn in the river. Did the ovary, then, as in the case of its fats and proteids, obtain the whole of the iron required from the muscle? or did part only come from muscle, and the remainder from some other source, *e.g.*, liver or blood?

To elucidate this the amount of iron in the muscles, ovaries, and in the livers of two typical fish leaving the sea in May, and in two typical fish from the upper reaches of the river in October, was determined. The very tedious nature of the analyses prevented a larger series of observations being accomplished. The close correspondence in the results obtained, however, indicates that the conclusions may be safely accepted.

Method.—The method employed in the analyses was that devised by Stockman (Journal of Phys., xviii., p. 485, 1895). The steps may, for convenience, be briefly recapitulated here.

Portions of the dried residue of the tissue, after extraction with ether, were dried in the hot-air chamber and weighed. In the case of the livers, a portion of the organ was pounded down and extracted with alcohol for several days, and then dried and weighed. They were then completely ashed in porcelain. The ash was extracted overnight with strong HCl, and next morning dilute H_2SO_4 was added, and the whole heated. It was then filtered through ash-free filter paper, and the filtrate placed on the steam bath to drive off the HCl. The iron was then left dissolved in the H_2SO_4. A few drops of potassium permanganate were added, and it was set aside for several days. If the colour remained (showing that all organic matter was destroyed), it was then reduced with zinc, and titrated with a standardised solution of potassium permanganate.

All the reagents, &c., were tested, and found iron-free.

The following Tables give the results of these analyses in grms. :—

TABLE I.

Per Cent. of Iron in Ovaries, Muscle, and Liver.

Estuary Fish.	Ovaries.	Muscle.		Liver.
		Thick.	Thin.	
20	0·003	0·0015	0·0018	·0160
25	0·004	0·0014	0·0018	—
Average,	0·0035	0·00145	0·0018	—

Upper-Water Fish.	Ovaries.	Muscle.		Liver.
		Thick.	Thin.	
63	0·003	0·002	0·001	·0180
69	0·002	0·002	0·001	—
Average,	0·0025	0·002	0·001	—

TABLE II.

Iron in Ovaries, Muscle, and Liver, in Fish of Standard Length.

Estuary Fish.	Ovaries.	Muscle.			Liver.
		Thick.	Thin.	Total.	
20	0·005	0·076	0·031	0·107	·0248
25	0·0059	0·0704	0·0308	0·109	—
Total,	0·0109	—	—	0·216	—
Average,	0·0054	—	—	0·108	—

Upper-Water Fish.	Ovaries.	Muscle.			Liver.
		Thick.	Thin.	Total.	
63	0·0631	0·067	0·014	0·081	·02489
69	0·0506	0·0603	0·0123	0·0726	—
Total,	0·1137	—	—	0·1536	—
Average,	0·0568	—	—	0·0763	—

From the above Table it will be observed that the ovaries gain in iron
at the expense of the muscle, thus:—

1. Loss of Iron.

Muscle of Fish in Lower Water $= 0\cdot1080$
Muscle of Fish in Upper Water $= 0\cdot0763$

Loss of Iron by Muscle $= 0\cdot0317$

2. Gain of Iron.

Ovaries of Fish in Upper Water $= 0\cdot0568$
Ovaries of Fish in Lower Water $= 0\cdot0054$

Gain of Iron by Ovaries $= 0\cdot0514$

Total Loss $= 0\cdot0317$
Total Gain $= 0\cdot0514$

Difference $= 0\cdot0197$

i.e., 39 per cent. of iron gained by the ovaries is not derived from the
iron stored in the muscles.

CONCLUSIONS.

It may be claimed that the results of the analyses have demonstrated (1) that the quantity of iron in the ovaries becomes distinctly
increased during the development of that organ, (2) that a considerable
amount of its iron is derived from the muscles, which become corre-
spondingly poorer in iron, (3) that none of it is derived from the liver.

Although the muscles supply the ovaries with the greater part of their
iron, yet there is a small quantity unaccounted for. The tissue which,
in all probability, supplies this, is the blood. If such observation had
been possible, it would have been of interest to investigate the changes
in the hæmoglobin of the blood. Since the iron has been shown to be
largely derived from the muscles, the amount from this source must be
comparatively small. There is no indication that the store of iron in
the liver is called upon. The iron which is stored up in the muscles
is probably obtained from the food, and kept in the muscles until the
development of the ovaries commences.

15.—THE PIGMENTS OF THE MUSCLE AND OVARY OF OF THE SALMON AND THEIR EXCHANGES.

By M. I. NEWBIGIN, B.Sc.

Among the many curious and interesting changes which the salmon undergoes throughout the year, not the least interesting is the variation in colour seen in the skin, the muscle, and the ovaries.

When the fish comes from the sea the skin is of a clear, bright silvery hue, while the flesh has the familiar strong pink colour. The small ovaries are of a yellow-brown colour. As the reproductive organs develop during the passage up the river, certain definite colour changes occur. The skin loses its bright silvery colour, and, more especially in the male, acquires a ruddy-brown hue. At the same time the flesh becomes paler and paler, and in the female the rapidly growing ovaries acquire a fine orange-red colour. The testes in the male remain a creamy white.

After spawning the skin tends in both sexes to lose its ruddy colour, and to regain the bright silvery tint; the flesh, however, remains pale until the kelt has revisited the sea. In other words, the salmon comes from the sea with a store of pigment in the muscles. During its sojourn in the river this pigment disappears from the muscles, is apparently in the female for the most part transferred to the ovaries, and so to the ova, and in both sexes is to a smaller extent deposited in the skin, there to undergo further changes. The accumulation of pigment in the muscle is associated with the presence of a large amount of fat, and fat and pigment disappear *pari passu*.

While in the Salmonidæ this colour change is most marked in the salmon, it is also observable in the sea trout. Even in certain varieties of brown trout, *e.g.* the Loch Leven trout, the pigmentation of the flesh is well marked when the fish are in prime condition, and becomes less marked as the genitalia develop. In all cases there seems to be the same close association between fat and pigment, and the simultaneous disappearance of the two.

Even outside the limits of the family of the Salmonidæ, pigmentation of the flesh is known to occur. Thus the Dawson salmon of the Australians (*Osteoglossum leichardti*), a member of a small tropical family, is described as having pink-coloured flesh, which tastes like that of the English salmon. The flesh of the Australian mud-fish (*Ceratodus forsteri*), again, is described as being oily and of a dark red or pink colour. Although there is no direct evidence, the descriptions would lead one to believe that in these cases also the pigment is associated with the presence of fat in the muscle.

As the pigments in these and other cases have been directly ascribed to the food, it was thought that their investigation in the case of the salmon would be of some interest. In this animal the pigments do not seem to have been previously studied.

Methods of Separation.

If the red flesh be pounded up in a mortar with sand, and then extracted with ether, it yields to the ether practically all its pigment. The ether becomes a deep golden-yellow colour, and leaves merely a greyish mass behind. On the evaporation of this extract a mass of pinkish pigment mixed with other substances is obtained. The pink pigment when treated with concentrated sulphuric or nitric acid gives a pure blue colour, which fades very rapidly, especially if water be added.

This reaction shows that the muscle contains one of the class of pigments called lipochromes, which are characterised by their solubility in ether, chloroform, benzol, petroleum ether, alcohol, etc.; by their colour, which varies from yellow to red ; and by the fact that they give a blue colour when treated with concentrated acid in the dry state. In order to obtain the pigment pure, for further study, the method of saponification was resorted to.

Two methods of saponification are available: (1) by means of metallic sodium in ethereal solution; (2) by means of caustic soda in alcoholic solution ; both methods were employed.

1. Small pieces of metallic sodium were added to a golden-yellow extract of the flesh in ether; on standing, the soap separated out at the base of the flask and was of a reddish colour, while the ether remained clear yellow. The ether was poured off and the soap washed with fresh ether, which did not extract the pigment. The soap was then dissolved in water, which became a pure pink colour. The addition of a little acetic acid to this solution gave a pink precipitate, which was filtered off. This pink precipitate dissolved readily in ether, to form, when dilute, a yellow solution, and in alcohol to form a pink solution. It also dissolved in petroleum ether, benzol, chloroform, etc., and gave a beautiful blue colour with concentrated sulphuric or nitric acid.

The yellow ether obtained by this method of saponification, after filtering off the soap, left on evaporation a pure yellow pigment which did not give a blue colour with concentrated sulphuric or nitric acids.

The ether of saponification is never anhydrous, and therefore when evaporated it usually leaves some drops of caustic soda solution behind , the yellow pigment readily dissolves in this.

2. In this method an alcoholic extract of the muscle was boiled with caustic soda. A slight precipitate of red pigment was obtained from the solution, but the mass of the pigment remained in solution with the soap.

To obtain the red lipochrome from this solution, two methods may be employed. The soap and the pigment may be precipitated by the addition of common salt, or the caustic solution may be shaken with ether in a separation funnel.

If excess of common salt be added to the caustic solution, a bright red soap comes down, leaving the solution a clear yellow colour. The soap may be washed with alcohol and then treated with dilute acid, after which the pigment is readily removed by alcohol, in which it forms a pink solution.

A simpler method is to shake the caustic solution after removal of the alcohol with ether, when the ether becomes deep yellow, leaving the caustic solution also yellow. The evaporation of the ethereal solution leaves a red pigment, which gives as before the lipochrome reactions.

The yellow pigment remains in the caustic solution, from which it cannot be extracted by ether or petroleum ether. It is not precipitated by the addition of acid, but the solution is then decolorized.

CHARACTERS OF PIGMENTS.

The above observations show that the red flesh of the salmon contains two pigments, of which one is pink, and gives the blue lipochrome reaction, while the other is yellow, and does not give this reaction.

An investigation of the mature ovaries conducted in precisely similar fashion showed that in them also two pigments—a red and a yellow—combine to produce the normal coloration of the organs.

The Red Pigment.—The characters of this red pigment are as follows: It is a lipochrome pigment, giving a blue colour in the dry state with nitric or sulphuric acid, and is soluble in alcohol, ether, benzol, petroleum ether, and acetic acid. Except in ether and petroleum ether, the solutions are of a pink or reddish colour, while these two solvents form pure yellow solutions. If the solutions are evaporated, the pigment recovers its red colour as the last drop of the solvent disappears.

The pigment forms compounds with caustic soda and potash, which are soluble in dilute alkaline solutions, at least in the presence of soaps. From these solutions the pigment may be precipitated by the addition of dilute acetic acid, or may be directly extracted by means of ether. Similar compounds are formed with lime and baryta.

The pure dry pigment fades very rapidly either in light or in darkness. Solutions also fade, but much more slowly. The loss of the power of giving a blue colour with concentrated acid is one of the first signs of change, and it may occur before the loss of colour in the solution is obvious. Pigment dissolved in benzol seems especially liable to undergo this change.

As to its other characters, solutions of the pigment when examined with a microspectroscope exhibit an indefinite shading in the neighbourhood of the F line, but this is hardly visible when the solutions are examined with a spectroscope of larger dispersion. In this case there is merely continuous absorption of the violet end.

As a whole, the pigment corresponds closely to the lipochrome pigment described in various animals, and notably in Crustacea, as tetronerythrin or zoonerythrin (by Moseley as crustaceorubrin). From the red pigment of the lobster the pigment differs slightly in tint and in the solubility of the sodium compound, but it is uncertain how much stress should be laid upon these differences.

The Yellow Pigment.—The yellow pigment does not give the lipochrome reaction. It belongs to a group of pigments which are apparently exceedingly widely distributed in the animal kingdom, but which have been little investigated. They have been commonly confounded with the lipochrome pigments. In the salmon the pigment occurs in the muscle, the ovary, and in large amount in the liver. It is always in close association with fat, and its solubility seems to depend upon that of the associated fat. In the salmon the pigment is associated with the fat olein, which is soluble in methylated spirit, and the pigment is also soluble in this solvent. In the case of bright yellow fat obtained from a cow, a pigment of otherwise identical characters was very little soluble in cold methylated spirit, but dissolved readily in ether. The fat with which the pigment was associated was here stearin and not olein.

The yellow pigment does not apparently form compounds with the alkalies or alkaline earths. It remains in the ether when an ethereal

L.

solution is saponified by metallic sodium, and in the caustic solution when an alcoholic solution is saponified by caustic soda.

DISTRIBUTION OF THE PIGMENTS.

It is interesting to note that the relative amounts of the two pigments vary considerably. As the flesh grows paler, it is the red pigment which seems to disappear first. While the ovary is small it contains chiefly the yellow pigment, while as it increases in size the amount of red pigment also increases. It would thus seem that there is a direct transference of the red pigment from the muscles to the ova, along with the transference of fat. Both pigments occur dissolved in fat.

SIGNIFICANCE OF THE PIGMENTS.

In this connection the first point of interest is that two pigments of similar or perhaps identical nature occur in the lobster and in all probability in other Crustacea. In the smaller Crustacea, as yet, the red pigment only has been described. In the Crustacea the pigments, especially the red one, are exceedingly important in producing the external coloration. In the lobster the digestive gland contains a considerable amount of fat which has a yellow pigment associated with it. The blood contains a red lipochrome, but no yellow pigment, while the hypodermis contains a large amount of red lipochrome, and apparently a small amount of yellow pigment. The red lipochrome forms a combination with some base, and then gives rise to the blue colour of the shell. (See Jour. of Physiol., 1897, p. 249). There is some reason to believe that in the lobster the yellow pigment and the red are closely related to one another. The yellow pigment contained in the digestive gland is in part got rid of by means of the alimentary canal, where it acts like a true bile-pigment in colouring the fæces, and is probably in part modified to form the red pigment of the hypodermis and shell.

Now, as to the relation of these facts to the pigmentation of the muscle of the salmon. The most obvious explanation is that the pigments of the salmon are derived directly from its food, and this is one which has been made by Günther, and accepted by other authors, e.g., by Beddard in his " Animal Coloration." At first sight the suggestion has much to recommend it. The pigments are very similar to those of the Crustacea, and perhaps identical with them. They disappear during the period of fasting, and are regained when the animal begins again to take food. In certain brown trout they appear sporadically, as if dependent upon particular diet, and finally the pigments are widely distributed in the Crustacea, occurring in one form or another in freshwater, littoral and abyssal forms.

There are, however, some difficulties in the way of the acceptance of this suggestion. In the first place, the salmon seems to feed chiefly on haddock, herring, and similar fish, so that the transfer of pigment can hardly be direct. The herring, however, feeds habitually on small Crustacea, so that it might be said that the pigments of the salmon are obtained indirectly from the herring which forms its food. In considering such a suggestion we have first to remember that the muscle contains two pigments, a red and a yellow, which simultaneously exist in the Crustacea. Of these pigments there is no reason to believe that the red exists in the herring. In three specimens examined I was unable to find any trace of it, either in the muscle or in the viscera. The stomachs in these cases were almost empty, but it hardly seems probable that the amount of pigment in the undigested food of the herring could be sufficient to supply all the colouring-matter of the salmon's muscle.

As to the yellow pigment, the viscera of the herring yielded to cold methylated spirit a small amount of a pigment which resembled the pigment obtained from the liver of the salmon, the same pigment being present in traces in a muscle extract. This suggests the possibility that the salmon obtains the yellow pigment of its muscle from food in association with fat, and that part of this pigment is modified to form the red. In the lobster there is some reason to believe that the yellow pigment is capable of being transformed into the red, and the conditions under which the two exist in the salmon suggest the possibility of a similar transformation there. As to the possibility of transference of yellow pigment from one organism to another, there is some evidence apart from the case of the salmon. Thus Poulton (Proc. Roy. Soc., London, liv., pp. 417-430; see also Nat. Sci., vol. viii., pp. 98-100) has shown by experiment that certain caterpillars derive their pigments from their food. Again, it is not uncommon to find the fat of sheep and cows dyed a deep yellow colour. According to some authorities, this occurs quite sporadically without known cause, while according to others special foods, notably maize, are the important agents. I have examined the yellow pigment of maize and compared it with pigment from yellow fat. The maize pigment gives the lipochrome reaction, faintly with sulphuric acid, distinctly with nitric, while the fat pigment gives no lipochrome reaction. In other respects, in tint, in solubility, and so on, the pigments closely resemble each other. This fact, taken in combination with Mr Poulton's experiments, seems to me at least to prove the possibility of the transference of these pigments from one organism to another, and therefore to suggest such an origin for the yellow pigment of the salmon.

This suggestion, however, gives rise at once to the difficulty that unless these three organisms can be shown to possess some common physiological peculiarity, then we are forced to the conclusion that all yellow pigments in animals are derived from their food—a conclusion for which there seems little evidence. Further, if the presence of pigment in the food is the only condition necessary to produce pigmented fat, it is difficult to understand why such coloured fat should not be universal in herbivorous animals, for all green parts of plants contain also a certain amount of yellow pigment.

It seems to me, however, that it is possible to point to a peculiarity possessed in common by the salmon, domesticated cattle, and caterpillars, namely, the habit in each case of ingesting food in excess of the normal requirements of the organism at the time of feeding. That this is so in the case of sheep and cattle undergoing the artificial process of fattening is obvious. Again, in both caterpillars and salmon the life-history is sharply divided into nutritive and reproductive periods, the periods occurring, respectively, once in the life of the caterpillar, and annually in the case of the salmon. During the nutritive period in both cases there is a large ingestion and deposition of fat, which later furnishes the energy used up during the reproductive period. It seems to me not unreasonable to suppose that while an organism which ingests a moderate amount of coloured fat is able to utilise or eliminate the pigment, and so deposit colourless fat in the tissues, one in which the ingestion of this coloured fat is excessive may be unable to do this, and so store coloured fat. In Poulton's caterpillars part of the pigment was eliminated with the faeces, which suggests that elimination is the natural fate of the pigment. If this explanation be correct, then it would follow that the reason for the coloration of the fat in sheep fed on maize must be that the ordinary diet contains as much pigment as it is possible for the organism to deal with, and the further addition of the pigmented fat of maize causes merely deposition of slightly altered pigment in the tissues.

If we apply this explanation to **the salmon, we have of course to face**
the possibility that both **the red and** yellow **pigment are** ingested in
this way with the food.   **On** the whole, this seems **to me** improbable,
and I am inclined to believe that it is only the yellow pigment **which is**
so obtained, **but that owing to** the conditions to **which** it is exposed in
the muscle, it becomes in part converted into **the red, which then gives**
rise to the colour of muscle and **ova.**

It is interesting to observe **that smolt which was kept at** Howietoun
for three years developed ripe ova.   **These had the characteristic red
colour.   (Day's** British and Irish Salmonidæ, p. 102).

I am thus of opinion that the **presence** of pigment-containing fat in
cattle, in caterpillars, and in **the salmon, is** due in **each case** to the habit
of ingesting coloured fatty **food in an amount which** is in excess of the
immediate requirements, the consequence being that fat coloured with
the pigment in a more or less modified condition is deposited in certain .
of the tissues.   While the pigment so deposited **is of** no importance in
cattle, in caterpillars **it** is important in producing the external coloration,
and in **the salmon in** colouring the ova.   In the male salmon the pigment is probably eliminated as the **fat** is used up.   The question is of
some interest, because if the suggestion here made be correct, it shows
that a characteristic pigmentation **may be** acquired as it were inciden-
tally in the course of the life history of the individual, under circumstances which render the question as to the inheritance of acquired
characters absolutely unimportant.

16.—ON THE CHANGES IN THE VALUE OF SALMON AS A FOOD STUFF.

By JAMES C. DUNLOP, M.D., F.R.C.P.E.

The measurements and analyses of salmon at the different seasons of the year, described in the previous sections, enable us to consider the value of the fish as a food stuff. Every salmon killed, whether early or late in the year, causes the loss of one breeding unit. If the stock is inexhaustible it matters not how many breeding units are destroyed. But if, as is the case, the stock is a diminishing one, economy demands that there must be a selection of fish, only those which give the best return being killed, and those which, on account of their poor value as a food stuff, do not compensate for that loss to the breeding stock, being preserved. The subject of killing salmon for sport is outside the scope of this section.

Two methods of considering the value of salmon flesh as a food stuff suggest themselves—(1) The value per unit of weight of flesh, per 100 grms. may be taken ; or (2) the total value per fish killed may be calculated. Both these will be considered.

The value of a food stuff is measured by the amount of energy which the combustion of its constituents produces, and is expressed as the number of calories—heat units—produced by that combustion. Calories are adopted as a convenient measurement of energy ; but when the energy value of a food stuff is expressed as so many calories, it is not implied that that food stuff is only capable of producing heat. One form of energy is capable of conversion into other forms, and so the measurement of the production of heat may be taken as a measurement of the energy available for all purposes. The calories referred to in this article are "great calories," each representing the amount of heat required to raise the temperature of 1000 cc. of water 1° Centigrade. For practical purposes the food value of the flesh of fish may be considered to depend on two constituents, proteids and fats, carbohydrates occurring in such a small amount that they may be disregarded. In the calculations which follow, the measurements and analyses of the fish are taken from the figures in the previous sections of this work, while the calorie value of proteid and fat is taken from Neumeister's Physiologische Chem., I., p. 282.

Food Value per Hundred Grms. of Salmon Flesh at Different Seasons.

In the following tables will be found a statement of the amount of proteid and fat and calorie equivalent of the flesh of the salmon at the different seasons, both from estuaries and upper waters. The calculations are based on the averages of the actual amounts found in each group of female fish received in 1896 (*vide* pp. 95 and 122).

Table I. gives the percentage of proteids and fats, and the food value in calories.

TABLE I.
ESTUARY FISH.
May and June.

	Thick Muscle.	Thin Muscle.	Total Muscle.
Proteid, . .	21·2	18·	20·6
Fat, . . .	10·2	17·9	12·1
Calories, . .	181·7	244·1	197·

July and August.

	Thick Muscle.	Thin Muscle.	Total Muscle.
Proteid, .	21·8	18·7	21·0
Fat, .	9·8	16·8	11·5
Calories, .	180·5	233·0	193·

October and November.

	Thick Muscle.	Thin Muscle.	Total Muscle.
Proteid, .	20·0	20·0	20·0
Fat, .	5·4	10·2	6·6
Calories, .	132·2	176·8	143·4

UPPER-WATER FISH.
May and June.

	Thick Muscle.	Thin Muscle.	Total Muscle.
Proteid, . .	21·2	20·6	21·0
Fat, . . .	6·8	10·3	7·7
Calories, . .	150·1	180·3	157·7

July and August.

	Thick Muscle.	Thin Muscle.	Total Muscle.
Proteid, . .	20·6	18·7	20·1
Fat, . . .	7·1	12·2	8·4
Calories, . .	150·5	190·1	160·5

TABLE I.—*Continued.*

October and November.

	Thick Muscle.	Thin Muscle.	Total Muscle.
Proteid,	17·5	16·2	17·2
Fat,	3·1	6·3	3·9
Calories,	100·6	125·0	106·7

TABLE II.

Showing Calories per 100 grms. of flesh of the various groups of fish.

	Estuary.	Upper Water.
May and June,	197·0	157·7
July and August,	193·0	160·5
October and November,	143·4	106·7

It will be seen in this:—

1. That the food value of salmon flesh is throughout the season greater when the fish is a fresh run fish, than when it has been some time in fresh water.

2. That the food value of salmon flesh diminishes as the season advances, both in estuary fish and upper-water fish.

3. That the flesh of salmon caught in the higher reaches of rivers, in October and November, is of much less value than other salmon flesh— being little more than half the value of the flesh of early estuary fish, and little more than two-thirds of that of earlier upper-water fish.

Food Value of Entire Fish at different Seasons.

The total food value of a salmon depends on the amount of flesh in the fish, and on the quality of that flesh. The amount of flesh possessed by a fish depends on the length of the fish and on its muscular development. These three factors and the resulting food values are shown in the following Table:—

TABLE III.

Showing Total Food Value of average fish of the various groups of fish, expressed as calories.

	Estuary.				Upper Water.			
	Length.	Total Weight of Muscle.	Calories per Cent.	Total Calories.	Length.	Total Weight of Muscle.	Calories per Cent.	Total Calories.
May and June,	75	2,667	197	5,254	75	2,477	158	3,800
July and August,	77	3,078	193	5,940	72	2,208	160	3,532
Oct. and Nov..	88	4,120	143	5,892	73	1,653	107	1,768

From this Table it will be seen:—

1. That the total food value of estuary fish remains nearly constant throughout the year, the poorer quality of the flesh in the later months being compensated for by the larger size of the average fish caught.

2. That from May to August the total food value of the fish caught in upper waters is about one-third less than that of estuary fish.

3. That the upper-water fish of October and November are of much less value as a food stuff than any other group of unspawned fish, their value being only one-half of that of the upper-water fish of the earlier months, and one-third of that of the estuary fish.

III.—SUMMARY OF RESULTS.

17.—GENERAL SUMMARY.

By D. NOËL PATON, M.D., F.R.C.P. ED.

A.—FACTORS DETERMINING MIGRATION.

It has been generally assumed that the passage of the salmon from the sea to the river is due to the *nisus generativus*. In considering the question it must be remembered that the Salmonidæ are originally fresh-water fish, and that the majority of the family spend their whole life in fresh water. Salmo Salar and other allied species have apparently acquired the habit of quitting their fresh-water home for the sea in search of food, just as the frog leaves the water for the same purpose. When, on the rich marine feeding grounds, as great a store of nourishment as the body can carry has been accumulated, the fish returns to its native element, and there peforms its reproductive act.

That the immigration of the fish is not governed by the growth of genitalia and by the *nisus generativus* is shown by the fact that salmon are ascending the rivers throughout the whole year with their genitalia in all stages of development (p. 64).

In fish leaving the sea the ovaries vary from 121 to 1439 grms. per fish of standard length, but the accumulation of solids in their muscles and ovaries together is about the same.

Solids in Muscle and Ovaries of Estuary Fish.

	November. (Winter Salmon.)	May and June.	July and Aug.	Oct. and Nov.
Muscle, - -	2481	2210	2270	1750
Ovaries,- -	23	47	72	545
Total, - -	2504	*2257	2342	2295

* Or including all fish dealt with on p. 83—
Muscle, ... 1990
Ovaries, ... 42
——
2032

The number of male fish examined was too small to allow of general conclusions being drawn.

In the kelts examined the amounts of solids were per standard fish —

Muscle,	-	-	-	946·00
Ovaries,	-	-	-	9·28

955·28

It would thus seem to be the *state of nutrition which is the factor determining migration towards the river; that when the salmon has accumulated the necessary supply of material it tends to return to its original habitat.*

From May to August—probably from November to August—the fish leaving the sea have the amount of material stored in their muscles about the same. During these months the ovaries are yet small, and do not act as a reservoir for stored material. In October and November the estuary fish have a smaller amount of stored material in their muscles. Why have these fish not left the sea sooner? Is it that, either because they have left the rivers later, or because the supply of food has been less readily obtained, the period of rapid growth of the genitalia has supervened before the full accumulation of material in the muscle has been accomplished? This rapid growth of the genitalia would withdraw material from the muscle and prevent its accumulating there, and thus, when the necessary amount of stored material was accumulated, it would be distributed between the muscles and genitalia.

The late-coming fish, although the supply of solids in the muscles is smaller, have the ovaries so large that the total store of nutrient material in the fish is just about the same as in those entering the estuaries in the earlier months.

A return to fresh water is essential for the completion of reproduction, for it has been shown that salt water prevents the development of the ova. In its natural condition the fish is impelled to migrate seaward in search of more abundant food, but descent to the sea is not necessary for the development of the genitalia. This is proved by the experiments carried on at Howietoun, which show that fish, when properly fed, may develop their genitalia without leaving fresh water. Again, the salmon of the great American lakes spawn in the streams, and yearly descend to the main lakes, as the salmon of this country descend to the sea, there to feed and lay in the necessary store of material.

The course of migration and the question of to-and-fro migration have been discussed on pp. 75 to 78, and it has been shown that the early-coming fish press up to the upper waters of the rivers, and that on to August fish continue to stream into the upper reaches; but that the fish leaving the sea in October and November do not at once ascend to the upper parts of the river. It has been further shown that there is strong evidence against there being a to-and-fro migration from river to sea and sea to river throughout the season.

B.—Do Salmon Feed in Fresh Water?

The question of whether salmon feed while in fresh water has been frequently discussed. Much depends on what is meant by the word "feeding." By *feeding*, we here mean not the mere swallowing of material; but the digestion, absorption, and utilisation of that material by the body. That salmon take the fly, minnow, or other shining object in the mouth is no argument as to their feeding in this sense. That they may, and occasionally do, take and swallow worms and other wriggling objects is well known. But the swallowing of a few worms can do but little to make good the enormous changes going on in the fish, even if, when swallowed, they are digested and used.

The evidence we have adduced may be summarised as follows:—

1st. There is no reason why salmon should feed during their stay in fresh water. When they leave the sea they have in their bodies a supply of nourishment not only sufficient to yield the material for the growth of ovaries and testes, but to afford an enormous supply of energy for the muscular work of ascending the stream (pp. 33, 93, and 120).

2nd. During the stay of the fish in fresh water the material accumulated in the muscles steadily diminishes, and there is absolutely no indication that its loss is made good by fresh material taken as food (pp. 83, 93, and 120).

3rd. The marked and peculiar degenerative changes which the lining membrane of the stomach and intestine undergoes during the stay of the fish in fresh water shows that during this period the organs of digestion are functionless (p. 13).

The absorption of food stuffs is not a mere mechanical process, but is chiefly dependent on the activity of the cells lining the alimentary canal, and in the river these essential cells degenerate and are shed.

It is a point of no little interest that before the fish again reaches the sea, after spawning, the lining membrane of the alimentary canal undergoes complete regeneration (p. 17 and 20), while the distended condition of the gall bladder seems to indicate that the bile-forming function of the liver is again becoming active.

4th. The very low digestive power of extracts of the mucous membrane of the stomach and intestine, not only in fish from the upper reaches in which the degenerative changes above referred to have occurred, but in fish coming to the mouth of the river and with the lining membrane still intact, seems to indicate that the salmon has practically ceased to feed before it makes for the river mouth (p. 23). It is to be regretted that, although every effort has been made, no specimen of a salmon stomach containing food has been procured. It is highly desirable that the digestive activity of such stomachs should be compared with the activity of those examined in this series of observations.

5th. The changes in the bacteria of the alimentary canal also throw light upon the question (p. 36). Generally speaking both in the estuary and in the upper reaches, the number of organisms varies directly with the temperature of the water. This is just what might be expected, since the number of organisms in the water largely depends upon its temperature.

But setting this aside, it is found that while in the gullet there is no great difference between the number of organisms in fish in the estuaries and in fish in the upper reaches, there is a markedly greater number of organisms in the stomach and intestine of fish in the upper reaches. This is especially the case with the putrefactive organisms which are the most readily destroyed by free acids. The greater abundance of these in the upper water fish is strongly indicative of the absence of the free acid which is formed in the stomach of fish while digesting food, and which if present would destroy them.

Miescher concluded that organisms are less numerous in the fish in the upper waters, but his conclusion is not supported by any evidence.

6th. Our observations confirm these of Miescher as regards the absence of food from the stomach. In not one of the 104 fish sent to the Laboratory during 1896 and the spring of 1897, was any trace of food found either in the stomach or in the contents of the intestine. That this is not due to rapid digestion has been proved. That it is not due to the fish disgorging the contents of the stomach when caught is shown by the absence of any trace of the indigestible portions of worms, insects, or fish in the intestine.

To the unscientific mind it is perhaps difficult to realise the possibility of a fast of several months in so active an animal as the salmon. But it must be remembered that it is simply a question of supply of energy.

The food yields energy for work, but if it is taken in excess, it is stored so as to be available at a future period. It has been shown that in the salmon such storage goes on to an enormous extent, and that, even at the end of the fast, there is still plenty of material available to meet any unexpected call for energy. Nor is the salmon exceptional in this respect. Many other cold-blooded animals have the same power of living for very prolonged periods without taking food, while several warm-blooded animals during the rutting season undergo prolonged fasts. It is stated that the male fur seal, after coming to land, may live for over a hundred days without food. During this period he is constantly engaged in struggles with other males, and he finally leaves the shore in a state of extreme emaciation.

We have thus no hesitation in confirming the conclusions of Miescher-Ruesch that the salmon, at least before spawning, does not feed during its sojourn in fresh water.

C.—Chemical Changes in the Salmon in Fresh Water.

It is because of this prolonged fast and because of the important changes going on in the fish during the fast that it affords so interesting a physiological study in metabolism. An opportunity is afforded of investigating the manner in which materials are stored in the animal body, the extent to which they may be transferred from one organ to another, the nature of some of the chemical changes they undergo, and the extent to which the various stored materials are utilised as a source of energy in the body.

I.—Solids and Water of Muscle, Genitalia, etc.

It has been shown that during the sojourn of the fish in fresh water there is a steady loss of solids from the muscles and a steady gain of solids by the genitalia, and it has further been shown that the gain of solids by the genitalia is small compared with the loss of solids from the muscle, that in fact the greater part of the solids lost from the muscles are used for some other purpose than the building up of the genitalia (p. 83).

As the season advances, the fish coming to the estuaries have a larger percentage of water in the muscle—about 6 or 7 per cent. more—than the fish leaving the sea earlier in the season. In the upper reaches the flesh throughout the season contains a greater percentage of water than the flesh of estuary fish. In May and June the upper-water fish have about 5 per cent. more water than the estuary fish, and in October and November about 13 per cent. more water than the estuary fish of May and June.

It is this increase in the percentage of water of the flesh which maintains the weight of the fish per fish of standard length, although the solids as a whole have diminished.

II.—Fats of Muscle, Genitalia, etc.

Nothing is more extraordinary than the enormous accumulation of fats which takes place in the muscle of the salmon during its visit to the sea (p. 95). Not only is the tissue between the individual fibres loaded with fat, but, as shown by Mr. Mahalanobis (p. 106), an intrafibrous or interfibrillar accumulation of fat occurs. In the river, as the season advances, this accumulated fat is steadily got rid of by the muscle. There is no reason to suppose that anything of the nature of a degeneration occurs. The fat is simply excreted from the muscle to supply the fat of the growing genitalia, or used in the muscle as a source of energy.

In the muscles the fatty acids are chiefly in the form of ordinary fats. In the ovaries and testes, on the other hand, the fatty acids are largely combined with phosphorus as lecithin. An important decomposition and reconstruction of the fats thus occurs in the growing ovaries. In the ovaries the amount of lecithin is very large, but the amount in the testes is by no means trifling, and the constant occurrence of this substance seems to point to it as the first stage in the formation of nucleins.

III.—Proteids of Muscle, Genitalia, etc.

Dr. Boyd's observations (p. 112) indicate that the albuminous materials of the muscle may be divided into two classes :—(1) Those soluble in salt solution ; (2) those not soluble in salt solution. He shows that globulin substances constitute nearly the whole of the soluble proteids, and that proteoses and peptones are not present in any circumstances. He further shows that there is a small quantity of some phosphorus-containing proteid—either a nuclein or a pseudo-nuclein—among the soluble proteids. It is these soluble proteids which diminish in fish in fresh water. When they are abundant, as in fish at the mouth of the river, on boiling they may coagulate between the flakes of the muscle and form with the fats the characteristic *curd.*

Of the insoluble proteids part is composed of white fibrous tissue, part of a phosphorus-containing proteid which may be called myostromin.

Dr. Dunlop's results (p. 120) show more fully the extent to which proteids accumulate in the muscles, and the rate at which they diminish as the fish passes up the river. The first point of interest is that the proteids do not disappear to anything like the same extent or at the same rate as the fats. As already indicated, it is from the fats that the energy for muscular work is chiefly procured. The second point of interest is that the proteid surplus available for energy—that is, the proteid not used in building the ovaries—is no greater in the upper water fish in October and November than in July and August. This seems to indicate that quite early in the season while the ovaries are growing slowly, the proteids disappearing from the muscle are more than sufficient to meet the requirements of these structures, while later in the year, when the growth of the ovaries is going on more rapidly, all the proteid disappearing from the muscle is transported to and used in them.

A further point of interest brought out is that in the male the amount of muscle proteid disappearing from the muscles is so much more than sufficient for the requirements of the growing testes that a very much larger surplus is available for muscular work in the male than in the female.

IV.—Source of the Energy for Muscular Work, etc. (p. 139).

The extent to which the fats and proteids lost from the muscles are used for the construction of the genitalia on the one hand, and for the liberation of energy on the other, varies somewhat in males and females. Taking the earlier months, to August, it has been shown that in the female 12 per cent. of the fats and 23 per cent. of the proteids go-to the ovaries, the rest being available for energy; while in the male about 5 per cent. of the fats and 14 per cent. of the proteids go to the testes.

The total energy liberated from fats and proteids is possibly somewhat greater in the male than in the female, being to August 1,271,000 Kgms. per fish of standard length in the female, and 1,380,000 Kgms. in the male. Of the energy thus liberated about 2,200 Kgms. are required to raise the fish to the height of the upper water of the river, the remainder being available for the much greater work of overcoming the resistance of the stream, for internal work and for other calls upon the energy supply.

Of this total available energy in the female, about 20 per cent. is derived from the proteids, while in the male only 9 per cent. is derived from this source. The rest is derived from the fats.

V.—*Phosphorus of Muscle, Genitalia, etc. (p. 143).*

It has been shown that in the female fish only just enough phosphorus is accumulated in the muscle to supply the wants of the growing ovaries, while in the male the accumulation is superabundant. In this connection it has been further pointed out that in the male the enormous growth of the bony jaw may use up a further amount of phosphorus. Whether in the female any phosphorus required for the ovaries in excess of that stored in the muscle is procured from the bones, these observations do not indicate.

The phosphorus is stored in the muscle chiefly as phosphates, and to a somewhat smaller extent as lecithin. The amount of lecithin in the muscle is not nearly sufficient to yield the lecithin of the ovaries. In the ovaries the phosphorus is in the form of ichthulin, a pseudo-nuclein and lecithin, so the phosphorus from the phosphates of the muscles must undergo profound changes in the growing ovaries, and being synthesised with organic bodies be built into these compounds. That these compounds are the forerunners of the still more complex nucleins of the embryo is indicated. In the male the transference of the phosphates of the muscle into these higher nuclein compounds is even more direct, and the occurrence of lecithin in considerable amount in the growing testes seems to point to this substance as the first step in the synthesis of inorganic phosphates to nucleic acid.

VI.—*Iron of Muscle and Ovaries.*

Dr Greig (p. 156) has shown that the ichthulin of the ovary contains iron, and the amount of iron in the ovaries thus increases as the organs grow. Whence is this iron procured?

It has been shown that the iron lost from the muscle is insufficient to yield the iron gained by the ovaries, and it is thus probable that the hæmoglobin of the blood must be drawn on for this element. The liver does not seem to yield iron to the ovaries.

VII.—*Pigments of Muscle and Ovaries.*

Miss Newbigin's study of the pigments of the muscle and ovary (p. 159) shows that two lipochromes are present. First, the very widely distributed yellow pigment—the so-called lutein which colours the yolk of the hen's egg; and second, a bright red lipochrome which, mixed with the former, gives the characteristic colour to the salmon muscle and ovaries.

Though it has not been possible to investigate the source of the pigments, the evidence adduced tends to show that the characteristic red pigment is probably not derived from the food, but that it is constructed possibly out of the very widely distributed yellow pigment. Its storage in the muscles and its transference to the ovaries has been demonstrated. Its fate in the male fish is still obscure, though the deeper pigmentation of the skin in the male suggests its elimination by that channel. What the purpose of the pigment is, is not clearly indicated, though it seems probable that by colouring the ova it may assist in their concealment during development.

VIII.—*Nature of the Transference of Material.*

On the nature of the transference of material these observations also throw important light. They clearly show that nothing of the nature of a degeneration in the muscle take place. The muscles simply excrete or give out the material accumulated in them.

Miescher discusses this point at great length. He first considers if what he calls the liquidation and degeneration (Fettentartung) is caused by changes in the nerves. But since he finds no visible changes in the nerve bundles to the muscles he dismisses this possibility. According to his view, the liquidation is caused by deficient respiration in the muscle, due to the deficient supply of blood as a result of the starvation and of the rapidly growing ovaries taking off a larger and larger amount of blood from the muscles. He supports this view first on general physiological principles, and secondly, from a consideration of the blood supply to the muscles and ovaries at various periods.

The theory, however, assumes that the change in the muscle is a degeneration, which it is not, and it affords at best but a partial explanation of the condition. It is too mechanical, and leaves unsolved the problem of what starts the growth of the ovaries, what causes the dilatation of blood-vessels there, and thus leads to the diminished blood supply to the muscles. The growth of the ovaries may be considered a cyclic function, but in all these cyclic functions the nervous system is intimately involved. It is well known that not only is the blood supply to every part of the body under the control of the nervous system, but that the very rate of chemical change in the cells of the body is also under the influence of the nerves. Not only does the brain bring about and control the extensive chemical changes in the muscles which lead to movements, but it also governs the slower chemical changes, such as those by which heat is produced in the warm-blooded animals. The building up and breaking down processes are alike controlled by nerves, and it is only fair to assume that the growth of the ovaries and testes and the discharge of material from the muscle for their growth are primarily determined by the nervous system, and that the vascular changes are secondary and not causal. In this connection it must be remembered that throughout the whole period the muscles remain active, and not only excrete material to the ovaries and testes, but also set free the energy of the proteids and fats stored within them, a state of matters irreconcilable with the idea of the existence of a degenerative process.

The investigation of kelts, though limited in extent, seems to show that, from the period of spawning to the return to the sea, the expenditure of energy is at a minimun. That many ova remain unshed in the abdomen is clearly shown, and that these ova are absorbed and shrivel up has also been proved. It is thus highly probable that, as indicated by Miescher-Ruesch, the kelt utilises its unshed spawn. We have discovered no evidence that kelts feed. In none of the 22 fish examined was there any trace of food in the stomach, or any remains of food in the intestine. On the other hand, a point of very great interest is the regeneration of the lining membrane of the stomach and intestine, and the reappearance of bile in the gall bladder, showing that the fish is again becoming capable of taking and using food.

IX.—Food Value of Salmon.

The food value per unit of weight of muscle deteriorates as the season advances. In each fish caught in the estuaries the food value remains almost constant, the larger size of the late-coming fish making up for the deterioration of the flesh. The food value of each fish caught in the upper waters is less than of those caught in the estuaries, and in October and November is only about one-third that of fish caught in the river-mouth.

This series of observations is only a contribution to a very large and very interesting subject. Many points yet remain to be investigated, while others touched upon here require extension and confirmation. As regards the course of migration, our investigations cover only a few months of the year, and interesting results are to be expected by extending the investigations into the other seasons.

We have given evidence to show that the early-coming fish occupy the upper reaches of the river, but a more extended series of investigations is required to show whether the late-coming fish, which during October and November are found in the lower waters, really spawn there, or whether they ultimately pass up to the upper waters.

Whether the *rate of migration* can be satisfactorily investigated in our short Scottish rivers is very doubtful. In the great Canadian rivers, such as the Fraser, very valuable results might be expected from the study of this question. Indeed it would be a matter of the greatest importance to have the observations recorded in these papers checked and extended on a large scale in such a river, with its unbounded supply of fish and hundreds of miles of water way.

The downward migration of kelts requires further study. Of the 22 kelts received in April, 1897, all were females. Is this a mere coincidence, or do the male kelts descend at a different time from the female? Some of the more important changes in the female kelts have been dealt with, but the interesting question of the loss of the great maxillary development in the male is yet to be elucidated.

The study of these and many other problems must be left for future investigations.

GLASGOW:

PRINTED BY JAMES HEDDERWICK & SONS,
FOR HER MAJESTY'S STATIONERY OFFICE.